Contributions to Management Science

The series *Contributions to Management Science* contains research publications in all fields of business and management science. These publications are primarily monographs and multiple author works containing new research results, and also feature selected conference-based publications are also considered. The focus of the series lies in presenting the development of latest theoretical and empirical research across different viewpoints.

This book series is indexed in Scopus.

Kang-Lin Peng · Yize Wu · Bo Gao

Space Tourism Supply Chain

Demand Research

Springer

Kang-Lin Peng
Faculty of International Tourism
and Management
City University of Macau
Macau, China

Yize Wu
Faculty of International Tourism
and Management
City University of Macau
Macau, China

Bo Gao
Faculty of International Tourism
and Management
City University of Macau
Macau, China

ISSN 1431-1941 ISSN 2197-716X (electronic)
Contributions to Management Science
ISBN 978-981-95-1079-5 ISBN 978-981-95-1080-1 (eBook)
https://doi.org/10.1007/978-981-95-1080-1

This Springer imprint is published by the registered company Springer Nature Singapore Pte Ltd.
The registered company address is: 152 Beach Road, #21-01/04 Gateway East, Singapore 189721, Singapore

Introduction

This monograph examines the space tourism value chain from the perspective of demand research. Demand is a crucial element in the formation of a market, and its economic behaviors exert a dynamic influence on tourism sciences, both in theoretical and practical contexts. This study emphasizes economic demand research within the monograph. We echo Song et al. (2023) on the research topic: Progress in tourism demand research: Theory and empirics in space tourism research. The research highlights advancements in general tourism studies through a comprehensive literature analysis. This monograph adheres to the established research framework but specifically addresses the emerging space tourism industry. The study identifies certain theoretical shortcomings in the methodologies applied to general tourism, which are deemed unsuitable for the study of space tourism. Furthermore, the final chapter elucidates the fundamental distinctions between space tourism and terrestrial tourism, proposing an innovative research trajectory for the exploration of space tourism (Peng et al., 2023).

The book addresses ten questions and one suggestion pertinent to research on space tourism demand. These include methodologies for measuring demand, identifying relevant variables, modeling nonlinearity, accounting for intertemporal dynamics, and evaluating the impact of COVID-19. The objective of these inquiries is to bridge theoretical and practical gaps within the field, as explored through research analyses and discussions in the book's chapters, which are briefly outlined in this Introduction as follows:

(1) How do we measure space tourism demand?

Measuring demand involves assessing market size, customer numbers, and revenue. Market reports, which project the market to reach USD 5.1 billion by 2035, driven by suborbital flights and lunar missions, indicating demand growth. Real cases include SpaceX's Inspiration4 mission, which sold tickets for USD 55 million each, providing data on willingness to pay (Lao Rosell, 2024).

(2) What are space tourism demand variables?

Demand variables include price, safety, risk perception, training required, launch location, and customer income. For example, a U.S. demand research highlights age, space tourism type (suborbital, orbital), and cost as key factors, with real-world examples like Virgin Galactic's suborbital flights targeting high-net-worth individuals (Edgell, 2024).

(3) How is nonlinearity in space tourism demand modeled?

Nonlinearity is modeled using advanced statistical methods like discrete choice experiments and machine learning. A study "Predicting Space Tourism Demand Using Explainable AI (Edgell, 2024)" uses a SpaceNet model to capture complex relationships, such as income and risk tolerance, with real cases like Blue Origin's pricing strategies showing nonlinear demand responses.

(4) How are intertemporal dynamics in space tourism demand accounted for?

Intertemporal dynamics examine the evolution of demand over time through the application of time series analysis or dynamic models. Although specific studies addressing the temporal changes in space tourism demand are limited, SpaceX's planned lunar missions serve as a case study illustrating future demand expectations within the framework of intertemporal dynamics (Zhang, 2024).

(5) What are the problems with studying space tourism demand forecasting with time series analysis?

Problems include limited historical data, non-stationarity, and external shocks. Given space tourism's nascency, traditional time series models struggle, a deep learning-based multivariate decomposition and ensemble framework for container throughput forecasting (Kulshrestha et al., 2024).

(6) Is effect size properly discussed in space tourism demand studies?

Effect size, measuring variable impact, is less discussed in space tourism research due to limited studies. We suggest a case study of SpaceX's revenue from Inspiration4 can adopt effect sizes in income-demand relationships study (Frantz Wilson & Henderson, 1991), though academic rigor is needed.

(7) Are descriptive statistics of the variables useful in space tourism demand modeling?

Descriptive statistics, like means and distributions, are useful for data exploration. Studies like "Modelling consumer choice behaviour in space tourism" use them to summarize customer characteristics, aiding models, with real data from Virgin Galactic's customer demographics as examples (Crouch et al., 2009).

(8) Which behavioral economics concepts have been applied in space tourism demand analyses?

Concepts like value-attitude-behavior theory and prospect theory are applied, as seen in "Psychology of space tourism marketing, technology, and sustainable development (Mehran et al., 2023)." Real cases, such as high-profile bookings by celebrities, reflect herd behavior and risk perception, influencing demand.

(9) How are macro and micro frameworks integrated in space tourism demand studies?

Integration involves combining macro factors (GDP, technology) with micro factors (individual preferences). The paper "Progress in tourism demand research: Theory and empirics (Song et al., 2023)" advocates this, with real cases like government support for SpaceX showing macro impacts and customer surveys reflecting micro choices.

(10) How did COVID-19 affect space tourism demand?

COVID-19 likely reduced demand due to travel restrictions and economic uncertainty. Real cases include delays in Virgin Galactic's flights, with recovery expected as seen in post-pandemic demand projections (Jensen & Sheller, 2023).

(11) Research issues and suggestions for space tourism demand research

This section, as outlined, will discuss current research gaps and suggest future directions for space tourism demand research (Peng et al., 2023), such as addressing data scarcity and methodological innovations, with real cases like SpaceX's data-driven approaches providing insights for future studies.

References

Crouch, G. I., Devinney, T. M., Louviere, J. J., & Islam, T. (2009). Modelling consumer choice behaviour in space tourism. *Tourism Management*, *30*(3), 441–454. https://doi.org/10.1016/j.tourman.2008.07.003

Edgell, R. (2024). The grand creative challenge of commercializing interplanetary space: An empirical comparative taxonomy of organizing models. In *Proceedings of the 75th International Astronautical Congress (IAC)*.

Frantz Wilson, M., & Henderson, M. (1991). Introduction to methodologic issues in minority research. Primary care research: Theory and methods—Conference proceedings.

Jensen, O. B., & Sheller, M. (2023). The future of aeromobilities in the light of global disruption. *Innovation: The European Journal of Social Science Research*, 1–21. https://doi.org/10.1080/13511610.2023.2182218

Kulshrestha, A., Yadav, A., Sharma, H., & Suman, S. (2024). A deep learning-based multivariate decomposition and ensemble framework for container throughput forecasting. *Journal of Forecasting*, *43*(7), 2685–2704. https://doi.org/10.1002/for.3151

Lao Rosell, G. (2024). *Study of commercial reusable launch vehicles business model: Case study of New Glenn from Blue Origin* [Universitat Politècnica de Catalunya]. http://hdl.handle.net/2117/403049

Mehran, J., Olya, H., & Han, H. (2023). Psychology of space tourism marketing, technology, and sustainable development: From a literature review to an integrative framework. *Psychology & Marketing*, *40*(6), 1130–1151. https://doi.org/10.1002/mar.21795

Peng, K.-L., Hsu, C. H. C., Lin, P. M. C., & Su, M. (2023). Proposing spacetime scale for space tourism economics. *Tourism Economics*, *29*(6), 1671–1678. https://doi.org/10.1177/13548166221109666

Song, H., Qiu, R. T. R., & Park, J. (2023). Progress in tourism demand research: Theory and empirics. *Tourism Management*, *94*, 104655. https://doi.org/10.1016/j.tourman.2022.104655

Zhang, Y. (2024). Space tourism: A new frontier of tourism experience. In *Handbook of experience science* (pp. 237–248). Edward Elgar Publishing. https://doi.org/10.4337/9781803926902.00025

Contents

List of Figures

Chapter 9

Chapter 10

Chapter 11

List of Tables

Chapter 11

Chapter 1
How to Measure Space Tourism Demand?

1 Introduction: Market Demand

Space tourism is a rapidly developing industry that can change the entire tourism sector. It has sparked the imagination of people worldwide, allowing them to traverse the Earth's atmosphere and experience zero gravity. Commercial space tourism involves individuals paying to travel to space for recreational purposes and participate in activities previously limited to astronauts (Chang, 2020). Understanding consumer demand is critical to its sustainability, as the sector hinges on accurately quantifying willingness-to-pay (WTP) and identifying key motivators beyond mere novelty-seeking. Early studies reveal that 58% of potential tourists prioritize "experiencing weightlessness" as the primary motivation, while 32% emphasize the symbolic value of "viewing Earth from space" as a transformative psychological event (Esken & Weinzimmer, 2022). By 2030, about 18–26% of market demand will come from ultra-high-net-worth individuals, though mass-market adoption requires addressing psychological barriers like risk perception—43% of surveyed respondents cite safety concerns as their top reservation (Collins & Autino, 2010; Olya & Han, 2020).

The willingness to pay (WTP) for space tourism may be significantly differentiated across generations. Younger generations (18–35 years old) show a higher willingness to pay, a tendency that is closely related to the consumer mentality of "prioritizing life experiences", a tendency that is further reinforced by the portrayal of adventurous behavior in social media (De Pelsmacker et al., 2005; Wey Smola & Sutton, 2002). Gender differences are also significant: men are more likely to associate space travel with "technical achievement" (68% of respondents), while women are more likely to view it as an "opportunity for personal growth" (61% of respondents). This difference in perception is consistent with the mechanism of gender differentiation in risk perception (Gnerre et al., 2022). Collectivist cultures prefer family or group participation in space experiences, while individualist cultures emphasise personal breakthroughs (Cohen, 1979). This difference echoes findings from cultural heritage tourism: collectivist societies tend to reinforce group identity through shared

K. Peng et al., *Space Tourism Supply Chain*, Contributions to Management Science,
https://doi.org/10.1007/978-981-95-1080-1_1

activities, while individualist societies are more concerned with personal narratives (Aranburu et al., 2016).

While social media amplifies public engagement, its measurement biases require caution (Hays et al., 2013). 78% of positive sentiment originates from non-paying spectators ("bystander effect"), creating a chasm between online enthusiasm and actual purchasing intent (Cox et al., 2009). VR-enhanced discrete choice experiments demonstrate a 22% increase in participants' WTP (Azizah & Cahyadi, 2023). 29% of environmentally conscious consumers reject space tourism despite affordability (Baciu, 2022), while "space justice" advocates condemn its role in exacerbating resource inequality (Marsh, 2024).

A synthesis of existing research reveals that the demand for space tourism is influenced by multiple factors, including technological advancements, cultural perceptions, and psychological motivations. While scholars have identified differences in willingness-to-pay across age and gender groups, and have highlighted how social media both drives demand and potentially exaggerates actual interest, significant research gaps remain: First, there are inconsistencies between survey results and findings from brain science experiments, and a lack of long-term data from actual space tourists to validate these findings. Second, environmental concerns and issues of social equity create conflicting emotions among potential consumers, an ethical dilemma rarely seen in traditional tourism. To address these gaps, this chapter will review the literature through three lenses: (1) comparing differences in measurement results from various research method; (2) analyzing how cultural differences across countries shape perceptions of space travel; and (3) exploring how price, safety, and environmental standards constrain market growth. These analyses will help establish a more comprehensive framework for assessing demand, providing insights not only for corporate market strategies but also addressing societal concerns about the sustainable development of space tourism. As commercial space technology matures, it is crucial for academia to deepen its understanding of the nature of demand to keep pace with the industry's rapid growth.

2 Literature Review: Space Tourism Studies

Maslow's hierarchy of needs theory divides human needs into five levels: physiological needs, safety needs, social needs, esteem needs, and self-actualization needs. Space tourism can meet people's needs at various levels. After meeting basic survival needs, space tourism can satisfy more than just esteem needs. Through space tourism, people can showcase their courage, abilities, and resources and gain recognition and appreciation from others. Additionally, space tourism can help people gain knowledge and skills, further satisfying their needs for respect and achievement and even rising to self-actualization needs. Space tourism provides a unique opportunity for people to explore the universe and learn more about Earth and the cosmos. This can help people realize their potential and satisfy their desire for self-actualization (Maslow, 1943).

Space tourism is a rapidly developing emerging industry, defined as commercial space travel, that aims to provide travelers with unique experiences such as leisure, entertainment, weightlessness, and astronomical observation (Spector & Higham, 2019). The industry can be further divided into suborbital, orbital, and Super-orbital tourism, depending on the orbit the spacecraft reaches (Cohen & Spector, 2020). Companies such as Virgin Galactic, Space Adventures, and SpaceX are essential participants in the industry, and governments worldwide are actively supporting financial projects related to space tourism (Chang, 2015).

Factors that influence potential customers for space tourism include nationality, education, gender, age, occupation, adventurous spirit, income, and personal preferences. As space travel is a nascent industry with limited existing data and customers to reference, Most studies focused on the potential customers of space travel. However, potential customers have similarities and differences. Some scholars argue that space travel is even more expensive than other adventure activities. For example, climbing Mount Everest costs nearly $1 million (Buhalis et al., 2006). Despite the high cost, other adventure activities, such as viewing auroras in the Arctic and Antarctic and exploring the Amazon, still have high customer demand. These customers are willing to pay extra to experience something unique and different. Therefore, potential customers of space travel can be broadly divided into two categories: wealthy leisure travelers and extreme sports enthusiasts (Olya & Han, 2020).

The imperative of gauging space tourism demand emanates from the inherent costliness and resource-intensive nature of space tourism endeavors. Given the substantial investments and infrastructural prerequisites entailed, a comprehensive understanding of the market demand for space tourism becomes imperative in assessing the viability and feasibility of commercializing space travel. Should the demand prove inadequate, the realization of a commercially sustainable space tourism industry could be jeopardized, plummeting the prospects of its progressive evolution. Consequently, elucidating the dimensions and trajectories of space tourism demand assumes pivotal significance in steering the longitudinal sustainability and evolutionary trajectory of the space tourism domain.

Moreover, the discernment of space tourism demand holds profound ramifications in delineating the contours of tailored marketing strategies aimed at effectively engendering heightened consumer engagement with space tourism experiences. By unraveling the intricacies of demand dynamics, the industry stands empowered to calibrate its promotional endeavors, market positioning, and consumer outreach initiatives to resonate more cogently with prospective space tourists. This, in turn, can catalyze a virtuous cycle of bolstered consumer interest, heightened participation, and sustained market growth, thereby underpinning the buoyancy and dynamism of the space tourism sector. Therefore, delving into the realm of space tourism demand not only furnishes critical insights for industry stakeholders to map out sustainable trajectories but also serves as a linchpin for igniting and fostering robust consumer traction within the ambit of space tourism. To better understand the potential customers of space travel, this study explores the impact of the above factors on space travel customers and proposes the following questions:

(1) How can we measure the demand for space tourism on a macro level?
(2) How can we measure the demand for space tourism on a micro level?

3 Methodology: Literature Review Analysis

The comprehensive exploration of space tourism demand is pivotal in elucidating the evolving landscape of commercial space exploration. This study employs a robust literature research method to delve into the multifaceted dimensions of space tourism demand, spanning macro and micro levels. By scrutinizing a spectrum of scholarly works, industry reports, and market analyses, this endeavor is poised to not only enrich and expand the existing body of literature on space tourism demand but also to furnish valuable insights that bear direct relevance to the actual development and commercialization of the burgeoning space tourism market. The in-depth elucidation of space tourism demand is thus geared towards not only sharpening the theoretical underpinnings of space tourism but also catalyzing practical implications that can actively steer and mold the trajectory of the commercial space tourism industry.

The Web of Science (WOS) Core Collection is a representative dataset for retrieving space tourism literature. We obtained a comprehensive list of space tourism papers from WOS using various keyword combinations. After reading abstracts and some widely cited articles and conducting multiple iterations of testing, we designed the following keyword combinations as our search criteria: "Space tourism" and "Suborbital space tourism." We collected space tourism papers published between January 1, 2010, and December 30, 2023, spanning 13 years, and 297 papers were retrieved. However, since many papers were unrelated to space tourism, we performed data cleaning. First, we only retained journal articles (67 papers were removed) and research papers based on WOS document types. We further refined the social science dataset, resulting in 72 papers. We then excluded papers not related to space tourism and narrowed the search results to 27 articles. Finally, we read all abstracts of this study (and if necessary, we read the complete related papers) to investigate papers related to space tourism in this topic, select tourism journals, and manually remove irrelevant papers. This led to a final dataset of 16 papers that met space tourism search criteria (Fig. 1).

The PRISMA framework (Preferred Reporting Items for Systematic Reviews and Meta-Analyses) for systematic literature reviews significantly improves the transparency and reproducibility of research through its standardised 27-item checklist and four-stage flowchart (Liberati et al., 2009). The framework requires researchers to clearly report key aspects of the literature screening process, risk of bias assessment, and data integration methods, thereby reducing selective reporting bias and enhancing the credibility of conclusions (Page et al., 2021). For example, its flowchart visually demonstrates the literature inclusion/exclusion process so that readers can visually verify the rigour of the study. The flowchart is a visualisation of the literature inclusion/exclusion process.

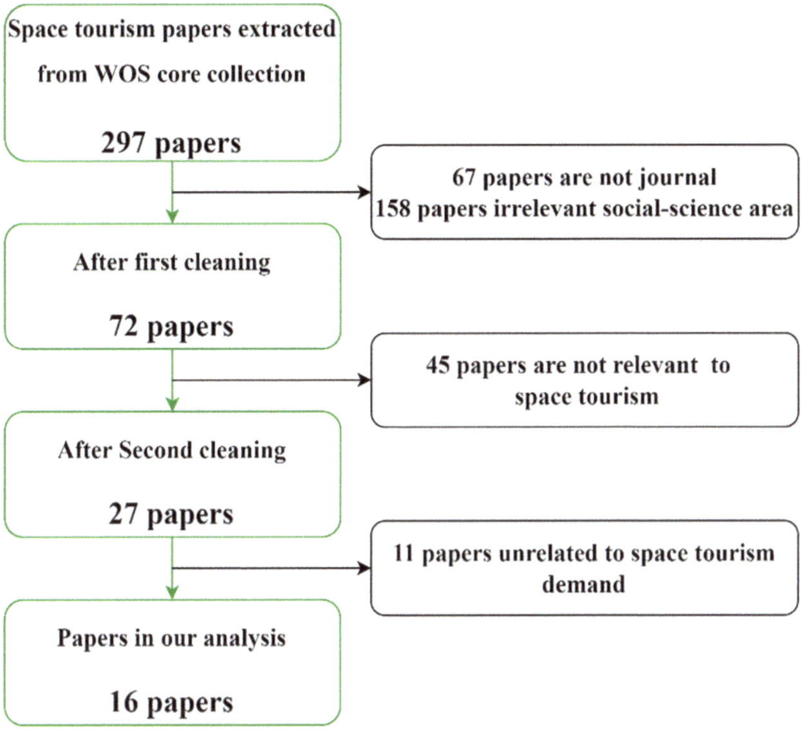

Fig. 1 Illustrates how we collected papers from this study

Space tourism, as an emerging cross-cutting field, has only accumulated about 109 core papers globally in the last 20 years (as of 2020). 27 If covers the main theoretical models and key empirical studies (North American vs. Asian demand) in the field, it is already representative. Systematic reviews emphasise the quality rather than the absolute number of included studies (Simunovic et al., 2009). The selection of studies follows strict PRISMA criteria allows for the development of a valid chain of evidence even if the total amount of literature is small. In qualitative systematic reviews, theoretical saturation is reached when additional literature fails to provide new themes (Butler et al., 2016). The current core dimensions of space tourism demand research (such as technology acceptance, pricing sensitivity) have achieved conceptual coverage through 27 articles. Combining the stage of development of the field with the requirements of the PRISMA methodology, the 27 articles are of sufficient analytical value to ensure screening rigour.

4 Results

Space Tourism Demands Main Path Analysis

Main path analysis is a powerful tool that can identify the chains of essential links in non-cyclic directed networks, thus extracting the skeleton of large complex directed networks (Liu & Kuan, 2016). Simplifying the network reveals the essential knowledge flows in citation networks and tracks the development path of research areas. Its advantage lies in simultaneously considering articles' direct and indirect effects, emphasizing the connections between citing and cited articles. Therefore, compared with the traditional "citation count" method that only considers direct effects, main path analysis reveals more accurate knowledge dissemination paths (Liu et al., 2013).

To track the evolution of Space Tourism Demand (STD) over time, we utilized the SPC algorithm to identify the primary citation path of an 8-article network. We comprehensively analyzed global and local main paths using this data, yielding consistent results. Figure 4, created using the "Pajek program package", showcases the global main path of STD research. The directional arrows in the figure represent the flow of knowledge, while the line thickness indicates the value of STD. The global main path of STD research is composed of 16 articles published between 2010 and 2023 (refer to Table 1). By analyzing these articles, we roughly divided STD research into two phases: an initial germination stage (2010–2016) and a phase of rapid development (2017–2023). Figure 2 shows the result.

Visual Analysis

Co-citation refers to two or more works cited by other literature simultaneously (Small, 1973), and co-citation frequency is the strength of co-citation between two cited works. Co-citation analysis has two main types: document and author co-citation, widely used in bibliometric analysis (Chen et al., 2010). Papers with the same cited references usually have similarities in concepts, research methods, or research topics (Osareh, 1996). Therefore, co-citation clusters formed by co-citation analysis help define the knowledge structure of a given field. The application of co-citation analysis continues to expand, and the research volume in information science is also growing rapidly. This study used Cite-Space software to conduct cluster analysis on the STD research's co-citation reference network to identify the current literature's knowledge structure.

From the perspective of journal visualization analysis, as indicated in Fig. 3, over the past 13 years, although space tourism belongs to the tourism industry, most of the articles are still published in space-focused journals, while the tourism journals have relatively less coverage with fewer related articles. Similarly, the journals related to measuring space tourism demand are also mainly space-focused journals, as seen in Fig. 4, where out of the 15 articles related to space tourism demand after the

Table 1 Variables at the macro and micro level

Measure variable	Details of measure variable	Paper	Number of papers
Variables at the micro level:			
Demographic factors	Age, gender, income, education, and other demographic factors can affect interest in and willingness to pay for a space tourism experience	Reddy et al. (2012); Wang et al. (2021); Zhang and Wang (2022); Mesa-Arango et al. (2023)	4
Attitudes and perceptions	Attitudes toward space tourism, perceptions of risk and safety, and other psychological factors affect demand	Reddy et al. (2012); Platt et al. (2020); (Olya & Han, 2020); Platt et al. (2020); Zhang and Wang (2022); Mesa-Arango et al. (2023)	6
Personal interests and preferences	Individual interests and preferences in space-related activities, such as stargazing or space exploration, can also affect demand	Reddy et al. (2012); Chang (2017);Zhang and Wang (2022); Mesa-Arango et al. (2023)	4
Availability of space tourism experiences	The availability and accessibility of space tourism experiences will also affect demand, such as the number of providers, the range of experiences offered, and the pricing of those experiences	Reddy et al. (2012); Le Goff and Moreau (2013); Zhang and Wang (2022); Mesa-Arango et al. (2023)	4
Marketing and promotion	Marketing and promotion of the space tourism experience will also affect demand, such as the type of information used and the platforms on which the marketing takes place	Orme (2017)	1
Variables at the macro level:			
Economic factors	Economic conditions, income levels, and disposable income can affect the demand for space tourism	Reddy et al. (2012); Webber (2013); Collins and Autino (2010);	3
Political factors	Government policy, international relations, and geopolitical stability can also affect the demand for space tourism	Reddy et al. (2012); Chang (2017)	2
Technological advances	Technological advances in space travel and exploration can affect demand for space tourism by making space tourism more accessible and affordable	Reddy et al. (2012)	1

(continued)

Table 1 (continued)

Measure variable	Details of measure variable	Paper	Number of papers
Cultural factors	National pride and interest in science and technology can also affect the demand for space tourism	Reddy et al. (2012); Collins and Autino (2010); Weibel (2020)	3
Environmental factors	Climate change and concerns about sustainability may affect the perceived value of space-related activities and thus the demand for space tourism	Reddy et al. (2012); Scott (2022); Collins and Autino (2010); Peeters (2018)	3

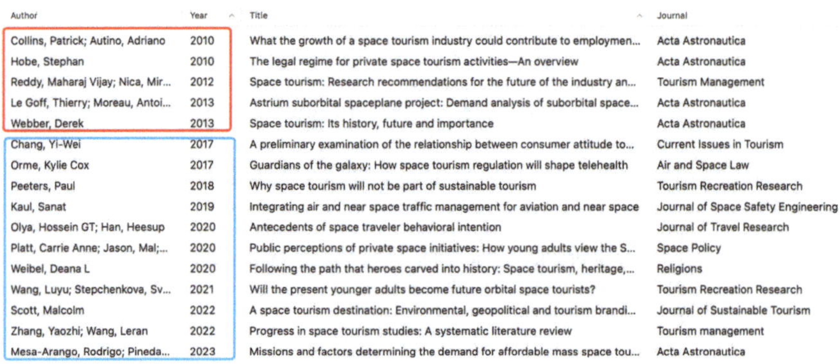

Author	Year	Title	Journal
Collins, Patrick; Autino, Adriano	2010	What the growth of a space tourism industry could contribute to employmen...	Acta Astronautica
Hobe, Stephan	2010	The legal regime for private space tourism activities—An overview	Acta Astronautica
Reddy, Maharaj Vijay; Nica, Mir...	2012	Space tourism: Research recommendations for the future of the industry an...	Tourism Management
Le Goff, Thierry; Moreau, Antoi...	2013	Astrium suborbital spaceplane project: Demand analysis of suborbital space...	Acta Astronautica
Webber, Derek	2013	Space tourism: Its history, future and importance	Acta Astronautica
Chang, Yi-Wei	2017	A preliminary examination of the relationship between consumer attitude to...	Current Issues in Tourism
Orme, Kylie Cox	2017	Guardians of the galaxy: How space tourism regulation will shape telehealth	Air and Space Law
Peeters, Paul	2018	Why space tourism will not be part of sustainable tourism	Tourism Recreation Research
Kaul, Sanat	2019	Integrating air and near space traffic management for aviation and near space	Journal of Space Safety Engineering
Olya, Hossein GT; Han, Heesup	2020	Antecedents of space traveler behavioral intention	Journal of Travel Research
Platt, Carrie Anne; Jason, Mal;...	2020	Public perceptions of private space initiatives: How young adults view the S...	Space Policy
Weibel, Deana L	2020	Following the path that heroes carved into history: Space tourism, heritage,...	Religions
Wang, Luyu; Stepchenkova, Sv...	2021	Will the present younger adults become future orbital space tourists?	Tourism Recreation Research
Scott, Malcolm	2022	A space tourism destination: Environmental, geopolitical and tourism brandi...	Journal of Sustainable Tourism
Zhang, Yaozhi; Wang, Leran	2022	Progress in space tourism studies: A systematic literature review	Tourism management
Mesa-Arango, Rodrigo; Pineda...	2023	Missions and factors determining the demand for affordable mass space tou...	Acta Astronautica

Fig. 2 Journal Main Path

final literature search, 8 are space-focused articles, accounting for approximately 53.33%. Among them, the journal that has published the most articles on space tourism demand is based on "Acta Astronautica," with 5 articles, accounting for 50% of the overall proportion. In tourism journals, such as "Tourism Management," only 2 articles (13.33%) exist. Although there are differences in the ratio of space and tourism-related articles in space and tourism-focused journals, it still indicates that space and tourism literature are the primary sources for studying space tourism demand literature. From the perspective of citation visualization analysis, it can be seen that there are not many articles studying space tourism demand in recent years, but it peaked in 2019, and the citation frequency has generally been on the rise and increasing year by year. The citation frequency reached a peak of 28 in 2020, as shown in Fig. 4. Overall, the number of articles related to space tourism demand in the development stage is more than in the embryonic stage, indicating that scholars are paying increasing attention to the issue of space tourism demand.

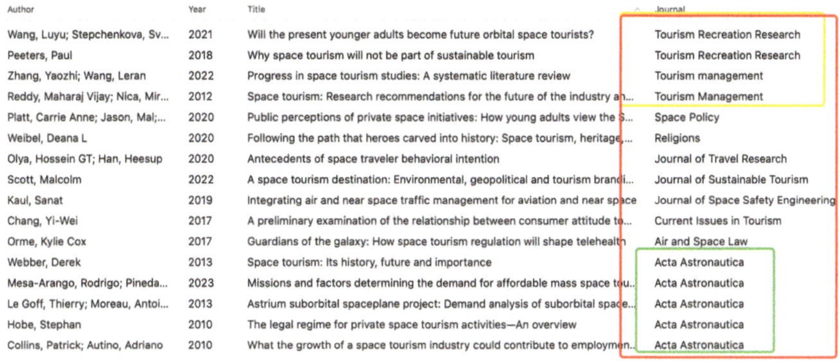

Author	Year	Title	Journal
Wang, Luyu; Stepchenkova, Sv...	2021	Will the present younger adults become future orbital space tourists?	Tourism Recreation Research
Peeters, Paul	2018	Why space tourism will not be part of sustainable tourism	Tourism Recreation Research
Zhang, Yaozhi; Wang, Leran	2022	Progress in space tourism studies: A systematic literature review	Tourism management
Reddy, Maharaj Vijay; Nica, Mir...	2012	Space tourism: Research recommendations for the future of the industry an...	Tourism Management
Platt, Carrie Anne; Jason, Mal;...	2020	Public perceptions of private space initiatives: How young adults view the S...	Space Policy
Weibel, Deana L	2020	Following the path that heroes carved into history: Space tourism, heritage,...	Religions
Olya, Hossein GT; Han, Heesup	2020	Antecedents of space traveler behavioral intention	Journal of Travel Research
Scott, Malcolm	2022	A space tourism destination: Environmental, geopolitical and tourism brandi...	Journal of Sustainable Tourism
Kaul, Sanat	2019	Integrating air and near space traffic management for aviation and near space	Journal of Space Safety Engineering
Chang, Yi-Wei	2017	A preliminary examination of the relationship between consumer attitude to...	Current Issues in Tourism
Orme, Kylie Cox	2017	Guardians of the galaxy: How space tourism regulation will shape telehealth	Air and Space Law
Webber, Derek	2013	Space tourism: Its history, future and importance	Acta Astronautica
Mesa-Arango, Rodrigo; Pineda...	2023	Missions and factors determining the demand for affordable mass space tou...	Acta Astronautica
Le Goff, Thierry; Moreau, Antoi...	2013	Astrium suborbital spaceplane project: Demand analysis of suborbital space...	Acta Astronautica
Hobe, Stephan	2010	The legal regime for private space tourism activities—An overview	Acta Astronautica
Collins, Patrick; Autino, Adriano	2010	What the growth of a space tourism industry could contribute to employmen...	Acta Astronautica

Fig. 3 Journal visualization

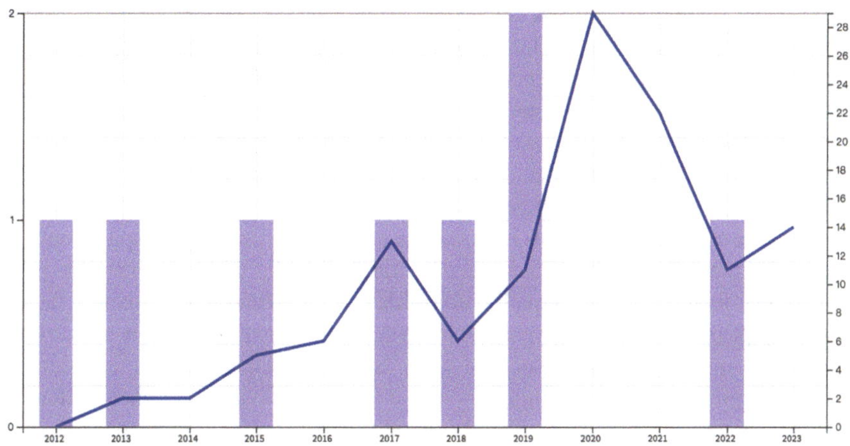

Fig. 4 Co-citation visualization (Light color: Journal, dark color: Citation frequency)

Micro and Macro Perspectives Analysis of STD

We have summarized the variables for measuring demand for space tourism from the literature, which can be classified into micro and macro factors, as shown in Table 1. Micro factors include demographic characteristics, attitudes and perceptions, personal interests and preferences, availability of space tourism experiences, and marketing and promotion. Demographic factors such as age, gender, income, and education can affect interest in and willingness to pay for space tourism. Attitudes and perceptions, such as attitudes toward space tourism and perceptions of risk and safety, as well as personal interests and preferences, can also impact demand. Availability of space tourism experiences, such as the number of providers and range of experiences offered, as well as marketing and promotion strategies, can also influence demand. There are several papers related to each of these micro factors, including Reddy et al.

(2012); Wang et al. (2021); Zhang and Wang (2022); Mesa-Arango et al. (2023); Orme (2017).

Macro factors, on the other hand, include economic, political, technological, cultural, and environmental factors. Economic factors such as economic conditions, income levels, and disposable income can affect demand for space tourism, as can political factors such as government policy and geopolitical stability. Technological advances in space travel and exploration can also impact demand by making space tourism more accessible and affordable. Cultural factors such as national pride and interest in science and technology can also influence demand. Lastly, environmental factors such as climate change and sustainability concerns may affect the perceived value of space-related activities and thus impact demand. There are several papers related to each of these macro factors, including Reddy et al. (2012); Webber (2013); Collins and Autino (2010); Chang (2017); Peeters (2018).

Based on Figs. 5 and 6 suggest that the literature primarily focuses on the micro level of demand measurement for space tourism. This means that studies are concerned with individual factors such as demographics, attitudes, perceptions, personal interests, and preferences. Among these factors, attitudes, interests, and personal payment ability are found to be the most significant determinants of demand for space tourism. On the macro level, studies look at broader factors such as economic conditions, political and cultural issues, technological advances, and environmental concerns. Among these macro variables, the economy, culture, and environment are considered the most important in shaping the demand for space tourism. In terms of macro space tourism demand, cultural variables (including national pride and interest in science) rank equally alongside economic and environmental factors as the top priorities for macro space tourism demand. However, there is scarce literature studying the cultural aspects of space tourism demand. The figures demonstrate the significance of examining micro and macro level variables when evaluating the demand for space tourism. This knowledge can be valuable for space tourism providers and policymakers as they formulate plans to meet consumer demand and foster sustainable growth in the industry.

Discussion

This study leveraging existing literature to comprehensively assess the demand for space travel. Through an extensive exploration and analysis of academic literature on space tourism demand within the Web of Science database, numerous articles focusing on micro-level perspectives regarding space tourism demand were identified. The wealth of literature presents a diverse array of insights, emphasizing the significance of examining both micro and macro-level factors when evaluating the demand for space tourism. By considering these multifaceted determinants, including individual motivations and broader market dynamics, a more nuanced understanding of space tourism demand can be attained.

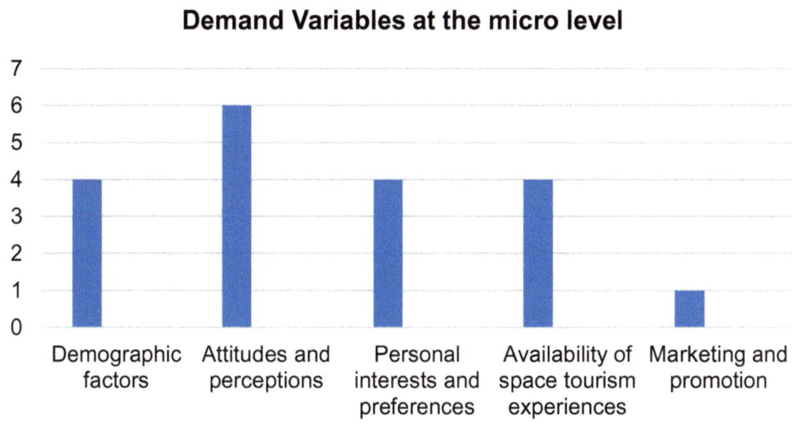

Fig. 5 Demand variables at micro level

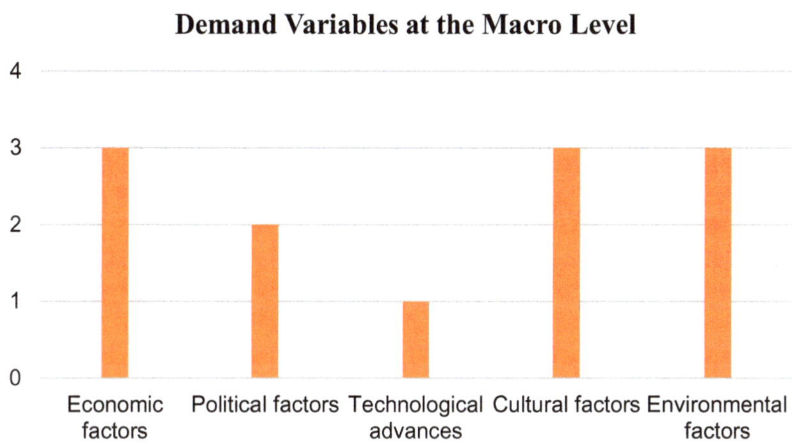

Fig. 6 Demand variables at macro level

There are significant differences in the measurement of space tourism demand across research methodologies, and this difference stems from the separation of methodological paths to explain consumer behavior. Quantitative Structural Equation Modeling (PLS-SEM) may overestimate the direct effect of exogenous variables (social status pursuit) and underestimate the moderating effect of interpersonal constraints (family opposition) when analyzing the relationship between motivation and behavioral intentions. Kim et al. (2024) found that the effect of interpersonal constraints on behavioral intentions for orbital/suborbital tourism was statistically insignificant ($p > 0.05$) when using PLS-SEM, but found that interpersonal constraints (attitudes of family and friends) were one of the necessary conditions to constitute high behavioral intentions through fuzzy set qualitative comparative analysis (fsQCA). This suggests that quantitative methods may ignore multifactor combinatorial effects, while qualitative methods are better at capturing nonlinear causal relationships. Experimental methods and questionnaires diverge in their measures of risk perception. The experimental method, by simulating a capsule environment (microclimate chamber experiment), found that visual and auditory multisensory stimulation reduced subjects' perception of physical risk (15% decrease in heart rate), whereas the questionnaire survey showed that respondents' concern for safety remained the primary constraint (Huang et al., 2023; Kim et al., 2024). This discrepancy reflects the separation between the effects of transient interventions on mental states in laboratory settings and rational assessments in real decision-making scenarios. Artificial Intelligence Forecasting Models (ANFIS) show higher sensitivity in demand forecasting. Hadavandi et al. (2011) combined feature selection with cluster analysis and found that the nonlinear relationship between price sensitivity and income level could not be identified in traditional regression models, whereas the ANFIS model reduces the prediction error by 22% through an adaptive neuro-fuzzy system (Kim et al., 2024). This shows the advantage of hybrid methods in capturing complex market dynamics.

National cultures influence demand generation paths by reconfiguring product meaning and risk frameworks. Individualistic cultures assign self-actualisation to space tourism, while collectivistic cultures rely more on group identity mechanisms for demand, and high-uncertainty avoidance cultures require authoritative validation to extend the trust-building cycle. Long-term orientated values drive demand towards an intergenerational educational function, with Laing and Frost (2019) finding that a much higher proportion of Chinese consumers (38%) than European and American consumers (12%) list 'space enlightenment for their children' as a motivation, while religious factors create market barriers through cosmological differences, with Schiavone (2018) confirms that fundamentalists are significantly more likely to reject space tourism (41%) than secular groups (9%), and that alien life hypotheses exacerbate this divide.

Price, safety and environmental protection are triple constraints: price sensitivity has a non-linear threshold, DePasquale et al. (2006) suggesting that price reductions of between $20 million and $12 million can lead to a seven-fold increase in demand elasticity, but that the weakening of luxury goods beyond the $8 million mark inhibits elasticity; safety has long-term negative impacts through the 'availability heuristic' (Weiss et al., 2001), and the difference in risk tolerance between Middle Eastern and Western groups increases market fragmentation. Safety has a long-term negative impact through the 'availability heuristic', with a single space accident lasting up to five years (Weiss et al., 2001) and market fragmentation fuelled by differences in risk tolerance between Middle Eastern and Western groups; environmental policy directly compresses profit margins, with the European Union's carbon trading system leading to a 60% increase in fares if it is implemented (Miraux et al., 2022), and the Nordic market Mandatory disclosure requirements for full-cycle carbon data on rocket fuel further elevate industry requirements.

The findings underscore the imperative of integrating various perspectives to form a comprehensive comprehension of space tourism demand. From individual consumer inclinations to overarching industry trends, the interplay between micro and macro-level factors elucidates the complex nature of space tourism demand. This holistic understanding is pivotal for stakeholders within the space tourism sector, providing them with essential insights to craft effective strategies that align with consumer demand and facilitate sustained, responsible growth within the industry.

For space tourism providers, these insights bear significant implications, guiding them in tailoring their offerings to meet the evolving demands and expectations of prospective space travelers. Likewise, policymakers can leverage this knowledge to formulate regulations and policies that support the burgeoning space tourism sector, fostering an environment conducive to its sustainable development. Ultimately, the amalgamation of micro and macro-level perspectives from the literature not only enriches our comprehension of space tourism demand but also offers actionable insights that can underpin informed decision-making and strategic planning within the industry.

5 Conclusion

As the space tourism industry evolves, measuring demand presents challenges and opportunities. Despite the potential benefits, such as new revenue streams and technological advances, one of the industry's primary challenges is the lack of historical data due to its relative newness. Developing accurate predictive models or conducting meaningful market analysis makes it challenging. To overcome this challenge, researchers must employ quantitative and qualitative methods to capture a range of variables related to individual attitudes, preferences, and behavior. This approach provides a comprehensive understanding of how residents make decisions regarding space tourism experiences and their resulting responses. Given the interdisciplinary and complex nature of space travel, input from experts with diverse backgrounds is necessary to achieve a nuanced understanding of demand. Looking at the subject from only one perspective or discipline can lead to one-sided and prejudiced predictions. Instead, stakeholders must accumulate knowledge and resolve disputes to understand the unique nature of space travel. While it is challenging to gather accurate data due to the nascent industry, the rapid development of science and technology provides opportunities to collect more real data in the future. Combined with empirical research and quantitative methods, this will enable stakeholders to explore new directions in consumer behavior and advance the industry further. As such, stakeholders must look forward to the future development of space travel and work collaboratively to overcome the challenges that lie ahead.

Understanding the demand for space tourism can provide insight into consumer behavior, preferences, and willingness to engage in novel experiences (Cater, 2010). Diverse academic analyses reveal the factors that influence individuals' decisions to participate in space tourism, providing valuable contributions to the field of consumer behavior. Measuring the demand for space tourism can help explore the economic potential of this emerging industry. It enriches academic work related to industrial economics and market analysis by enabling researchers to assess market size, pricing strategies and revenue forecasts (Nicolaisen & Driscoll, 2014). Demand measurement is critical to inform the policy and regulatory frameworks governing space tourism. This chapter contributes to the understanding of legal and ethical considerations associated with commercial space travel and contributes to a sound governance structure (Freeland, 2005).

Exploring the need for space tourism allows for philosophical inquiry into the relationship between humans and space exploration and the psychological motivations of individuals interested in adventures beyond Earth. This supports interdisciplinary scholarship at the intersection of philosophy, psychology, and space studies (Spector & Higham, 2019). Measuring the demand for space tourism provides scholars with the opportunity to critically assess the ethical and environmental implications of expanding commercial space travel. This contributes to ethical discourse in fields such as philosophy, ethics, and environmental studies (Chang, 2015). Studying the demand for space tourism from an academic perspective can contribute to the discussion of scientific advances, technological innovations and the integration of aerospace engineering with consumer needs, thus enriching academic engagement in areas such as engineering, physics and technology studies (Toivonen, 2022).

Future research in the realm of space tourism presents several promising avenues for exploration. Firstly, an in-depth analysis of cultural and national disparities in space tourism demand stands as a pertinent area for investigation. Understanding and delineating potential variations in demand among diverse cultures and nationalities is crucial, offering insights into the nuanced drivers of interest in space travel. By delving into these distinctions, researchers can uncover underlying factors that influence individuals' inclinations towards space tourism, potentially shedding light on cultural perceptions, values, and aspirations that shape demand. Secondly, an imperative focus of future research involves probing into the environmental ramifications of space tourism. This entails a comprehensive examination of the industry's ecological footprint, including considerations such as carbon emissions and the proliferation of space debris. Addressing these concerns necessitates an inquiry into potential mitigation strategies, with the goal of minimizing adverse impacts while fostering sustainable practices within the space tourism domain. Such investigations are essential for ensuring the responsible development of space tourism, aligning with broader sustainability objectives. Lastly, a compelling area for future inquiry lies in elucidating the role of social media and marketing mechanisms in shaping space tourism demand. Investigating the intricate interplay between digital platforms, marketing strategies, and consumer interest offers valuable opportunities to comprehend how perceptions and desires regarding space travel are influenced and cultivated. Identifying effective approaches to leverage social media and marketing dynamics to bolster interest in space tourism can be instrumental in expanding and diversifying the demographic appeal of this burgeoning sector.

Directed research efforts aimed at advancing the field of space tourism promise to yield a rich harvest of insight, thereby facilitating informed decision-making, promoting sustainable practices, and nurturing innovative strategies to buoy the industry towards enduring eminence and viability. Building upon this foundation, future research endeavors in space tourism could be strategically channeled into several core areas of inquiry. The first area of focus entails a nuanced analysis of cultural and national disparities in space tourism demand, with emphasis on discerning variations in consumer inclination across divergent cultural and national contexts. Secondly, an imperative facet for future exploration involves a comprehensive investigation into the environmental ramifications of space tourism, encompassing the assessment of carbon emissions and space debris, and proffering viable pathways to offset and abate these ecological impacts. Lastly, an incisive study delving into the interplay of social media and marketing in sculpting the contours of space tourism demand is warranted, coupled with an astute identification of strategic paradigms for harnessing these drivers to augment interest and engagement within the sphere of space tourism. These delineated research trajectories promise to engender substantial value in guiding and enriching the evolution of space tourism, redounding to a more holistic understanding of the multifaceted dynamics and imperatives underpinning this emergent domain.

References

Aranburu, I., Plaza, B., & Esteban, M. (2016). Sustainable cultural tourism in urban destinations: Does space matter? *Sustainability, 8*(8), 699. https://doi.org/10.3390/su8080699

Azizah, A. N., & Cahyadi, E. R. (2023). Preferensi dan Willingness To Pay Wisatawan Terhadap Virtual Reality di Malang Night Paradise Pascapandemi. *Jurnal Aplikasi Bisnis dan Manajemen (JABM), 9*(1), 261–261. https://doi.org/10.17358/jabm.9.1.261

Baciu, R. -G. (2022). Space tourism. A future sustainable business or not? *Cactus, 4*. https://doi.org/10.24818/CTS/4/2022/1.06

Buhalis, D., Costa, C., & Ford, F. (2006). *Tourism business frontiers*. Routledge.

Butler, A., Hall, H., & Copnell, B. (2016). A guide to writing a qualitative systematic review protocol to enhance evidence-based practice in nursing and health care. *Worldviews on Evidence-Based Nursing, 13*(3), 241–249. https://doi.org/10.1111/wvn.12134

Cater, C. I. (2010). Steps to space; opportunities for astrotourism. *Tourism Management, 31*(6), 838–845. https://doi.org/10.1016/j.tourman.2009.09.001

Chang, Y.-W. (2015). The first decade of commercial space tourism. *Acta Astronautica, 108*, 79–91. https://doi.org/10.1016/j.actaastro.2014.12.004

Chang, Y.-W. (2017). A preliminary examination of the relationship between consumer attitude towards space travel and the development of innovative space tourism technology. *Current Issues in Tourism, 20*(14), 1431–1453.

Chang, Y. -W. (2020). From aviation tourism to suborbital space tourism: A study on passenger screening and business opportunities. *Acta Astronautica, 177*, 410–420. https://doi.org/10.1016/j.actaastro.2020.07.020

Chen, C., Ibekwe-SanJuan, F., & Hou, J. (2010). The structure and dynamics of cocitation clusters: A multiple-perspective cocitation analysis. *Journal of the American Society for Information Science and Technology, 61*(7), 1386–1409.

Cohen, E. (1979). A Phenomenology of Tourist Experiences. *Sociology, 13*(2), 179–201. https://doi.org/10.1177/003803857901300203

Cohen, E., & Spector, S. (2020). Space tourism-past to future: A perspective article. *Tourism Review, 75*(1), 136–139. https://doi.org/10.1108/TR-03-2019-0083
Collins, P., & Autino, A. (2010). What the growth of a space tourism industry could contribute to employment, economic growth, environmental protection, education, culture and world peace. *Acta Astronautica, 66*(11–12), 1553–1562. https://doi.org/10.1016/j.actaastro.2009.09.012
Cox, C., Burgess, S., Sellitto, C., & Buultjens, J. (2009). The role of user-generated content in tourists' travel planning behavior. *Journal of Hospitality Marketing & Management, 18*(8), 743–764. https://doi.org/10.1080/19368620903235753
De Pelsmacker, P., Driesen, L., & Rayp, G. (2005). Do consumers care about ethics? Willingness to pay for fair-trade coffee. *Journal of Consumer Affairs, 39*(2), 363–385. https://doi.org/10.1111/j.1745-6606.2005.00019.x
DePasquale, D., Charania, A., & Olds, J. (2006). The emerging orbital space tourism industry: New insight into demand and prospects for success. In *Space 2006* (pp. 7478). https://doi.org/10.2514/6.2006-7478
Esken, C. A., & Weinzimmer, L. (2022). Riskier business: How consumer perceptions of corporate political advocacy have evolved. *Journal of Management Policy & Practice, 23*(3). https://doi.org/10.33423/jmpp.v23i3.5728
Freeland, S. (2005). *Up, up and... back: The emergence of space tourism and its impact on the international law of outer space* (Vol. 6).
Gnerre, M., Abati, D., Bina, M., Confalonieri, F., De Battisti, S., & Biassoni, F. (2022). Risk perception and travel satisfaction associated with the use of public transport in the time of COVID-19. The case of Turin, Italy. *PloS one, 17*(3), e0265245. https://doi.org/10.1371/journal.pone.0265245
Hadavandi, E., Ghanbari, A., Shahanaghi, K., & Abbasian-Naghneh, S. (2011). Tourist arrival forecasting by evolutionary fuzzy systems. *Tourism Management, 32*(5), 1196–1203. https://doi.org/10.1016/j.tourman.2010.09.015
Hays, S., Page, S. J., & Buhalis, D. (2013). Social media as a destination marketing tool: Its use by national tourism organisations. *Current Issues in Tourism, 16*(3), 211–239. https://doi.org/10.1080/13683500.2012.662215
Huang, X.-T., Wang, J., Wang, Z., Wang, L., & Cheng, C. (2023). Experimental study on the influence of virtual tourism spatial situation on the tourists' temperature comfort in the context of metaverse. *Frontiers in Psychology, 13*, 1062876. https://doi.org/10.3389/fpsyg.2022.1062876
Kim, M. J., Hall, C. M., Kwon, O., Hwang, K., & Kim, J. S. (2024). Orbital and sub-orbital space tourism: Motivation, constraint and artificial intelligence. *Tourism Review, 79*(2), 392–407. https://doi.org/10.1108/TR-01-2023-0017
Laing, J., & Frost, W. (2019). Exploring motivations of potential space tourists. In *Space Tourism* (Vol. 25, pp. 141–161). Emerald Publishing Limited. https://doi.org/10.1108/S1571-504320190000025008
Le Goff, T., & Moreau, A. (2013). Astrium suborbital spaceplane project: Demand analysis of suborbital space tourism. *Acta Astronautica, 92*(2), 144–149. https://doi.org/10.1016/j.actaastro.2013.03.025
Liberati, A., Altman, D. G., Tetzlaff, J., Mulrow, C., Gøtzsche, P. C., Ioannidis, J. P., Clarke, M., Devereaux, P. J., Kleijnen, J., & Moher, D. (2009). The PRISMA statement for reporting systematic reviews and meta-analyses of studies that evaluate healthcare interventions: explanation and elaboration. *Bmj, 339*. https://doi.org/10.1136/bmj.b2700
Liu, J. S., & Kuan, C. H. (2016). A new approach for main path analysis: Decay in knowledge diffusion. *Journal of the Association for Information Science and Technology, 67*(2), 465–476.
Liu, J. S., Lu, L. Y., Lu, W.-M., & Lin, B. J. (2013). Data envelopment analysis 1978–2010: A citation-based literature survey. *Omega, 41*(1), 3–15.
Marsh, M. S. (2024). *Ethical and medical dilemmas of extreme tourism*. International Conference on Tourism Research.
Maslow, A. H. (1943). A theory of human motivation. *Psychological Review, 50*(4), 370.

Mesa-Arango, R., Pineda-Jaramillo, J., Araujo, D. S., Bi, J., Basva, M., & Viti, F. (2023). Missions and factors determining the demand for affordable mass space tourism in the United States: A machine learning approach. *Acta Astronautica*. https://doi.org/10.1016/j.actaastro.2023.01.006

Miraux, L., Wilson, A. R., & Dominguez Calabuig, G. J. (2022). Environmental sustainability of future proposed space activities. *Acta Astronautica, 200*, 329–346. https://doi.org/10.1016/j.actaastro.2022.07.034

Nicolaisen, M. S., & Driscoll, P. A. (2014). Ex-post evaluations of demand forecast accuracy: A literature review. *Transport Reviews, 34*(4), 540–557. https://doi.org/10.1080/01441647.2014.926428

Olya, H. G. T., & Han, H. (2020). Antecedents of space traveler behavioral intention. *Journal of Travel Research, 59*(3), 528–544. https://doi.org/10.1177/0047287519841714

Orme, K. C. (2017). Guardians of the galaxy: How space tourism regulation will shape telehealth. *Air and Space Law, 42*(2).

Osareh, F. (1996). *Bibliometrics, citation analysis and co-citation analysis: A review of literature I*.

Page, M. J., Moher, D., Bossuyt, P. M., Boutron, I., Hoffmann, T. C., Mulrow, C. D., Shamseer, L., Tetzlaff, J. M., Akl, E. A., & Brennan, S. E. (2021). PRISMA 2020 explanation and elaboration: updated guidance and exemplars for reporting systematic reviews. *Bmj, 372*. https://doi.org/10.1136/bmj.n160

Peeters, P. (2018). Why space tourism will not be part of sustainable tourism. *Tourism Recreation Research, 43*(4), 540–543. https://doi.org/10.1080/02508281.2018.1511942

Platt, C. A., Jason, M., & Sullivan, C. J. (2020). Public perceptions of private space initiatives: How young adults view the SpaceX plan to colonize Mars. *Space Policy, 51*, Article 101358.

Reddy, M. V., Nica, M., & Wilkes, K. (2012). Space tourism: Research recommendations for the future of the industry and perspectives of potential participants. *Tourism Management, 33*(5), 1093–1102. https://doi.org/10.1016/j.tourman.2011.11.026

Schiavone, S. R. (2018). *A place among the stars? The influence of religion and creationism on attitudes towards space exploration and beliefs in extraterrestrial life*. https://doi.org/10.13023/etd.2018.276

Scott, M. (2022). A space tourism destination: Environmental, geopolitical and tourism branding considerations for New Zealand as a 'launch state.' *Journal of Sustainable Tourism, 30*(9), 2240–2253. https://doi.org/10.1080/09669582.2020.1817049

Simunovic, N., Sprague, S., & Bhandari, M. (2009). Methodological issues in systematic reviews and meta-analyses of observational studies in orthopaedic research. *JBJS*, *91*(Suppl. 3). https://journals.lww.com/jbjsjournal/fulltext/2009/05003/methodological_issues_in_systematic_reviews_and.14.aspx

Small, H. (1973). Co-citation in the scientific literature: A new measure of the relationship between two documents. *Journal of the American Society for Information Science, 24*(4), 265–269.

Spector, S., & Higham, J. E. S. (2019). Space tourism in the Anthropocene. *Annals of Tourism Research, 79*. https://doi.org/10.1016/j.annals.2019.102772

Toivonen, A. (2022). Sustainability dimensions in space tourism: The case of Finland. *Journal of Sustainable Tourism, 30*(9), 2223–2239. https://doi.org/10.1080/09669582.2020.1783276

Wang, L., Stepchenkova, S., & Kirilenko, A. P. (2021). Will the present younger adults become future orbital space tourists? *Tourism Recreation Research, 46*(1), 109–123.

Webber, D. (2013). Space tourism: Its history, future and importance. *Acta Astronautica, 92*(2), 138–143.

Weibel, D. L. (2020). Following the path that heroes carved into history: Space tourism, heritage, and faith in the future. *Religions, 11*(1), 23.

Weiss, K. A., Leveson, N., Lundqvist, K., Farid, N., & Stringfellow, M. (2001, October 14–18). *An analysis of causation in aerospace accidents*. 20th DASC. 20th Digital Avionics Systems Conference (Cat. No.01CH37219).

Wey Smola, K., & Sutton, C. D. (2002). Generational differences: Revisiting generational work values for the new millennium. *Journal of Organizational Behavior: THe International Journal of Industrial, Occupational and Organizational Psychology and Behavior, 23*(4), 363–382. https://doi.org/10.1002/job.147

Zhang, Y., & Wang, L. (2022). Progress in space tourism studies: A systematic literature review. *Tourism Management, 47*(4), 372–383.

Chapter 2
What Are Space Tourism Demand Variables?

1 Introduction

The human desire for the sky has not diminished since the beginning of time. Wan Hu of the Ming Dynasty once strapped gunpowder to a chair in the hope that he could use it to fly into the sky, but although it failed, he was described as the world's first "astronaut." The aviation period began in December 1903 with the growth and improvement of technology when the Wright Brothers successfully flew a heavier-than-air aircraft in Kitty Hawk, North Carolina, USA. The first human to travel to space was Yuri Gagarin of the Soviet Union in 1961.

Thus, we arrived in 2001, 98 years after the Wright Brothers' maiden flight and 40 years after Yuri Gagarin's first space journey, when Dennis Tito became the first advanced ape to travel into space. He spent seven days in orbit around the Earth on the International Space Station after paying $20 million to travel to space on a Soyuz spacecraft. It suggests that space travel is becoming a possibility.

Although space tourism has been around for a while, it has recently gained ground due to technological developments and the commercial space industry. Based on the spacecraft's trajectory and the unique experience it offers visitors, space tourism can be broadly divided into sub-orbital, orbital, and beyond-orbital tourism (Spector, 2020). This chapter focuses on sub-orbital and orbital; until 2022, 12 space tourism service providers will conduct space tourism services (Hurley, 2022).

From 2001 to 2021, the market size of sub-orbital and orbital space tourism was 22 customers, with a total cost estimated at $846 million. By 2021 it accounted for 64% of customers and 67% of expected sales. The market growth from 2021 to 2022 is expected to increase by 257% for private astronauts and 98% in sales (Hurley, 2022). With the growth of commercial space tourism over the last two decades, the size and demand of the market has been an inevitable topic of discussion. The market has great demand for sub-orbital and orbital space tourism. For the market, without demand, there is no revenue, and it cannot operate properly.

K. Peng et al., *Space Tourism Supply Chain*, Contributions to Management Science,
https://doi.org/10.1007/978-981-95-1080-1_2

Current research into tourism demand variables has focused on macro and micro aspects. At the macro level, the primary factors influencing tourism demand are GDP, currency rate, health risk, and relative pricing. The most crucial elements at the micro level include money, cost, age, and traveling companion. Due to its unique qualities, it can alter slightly in the context of space tourism.

Specifically, The growth trajectory of space tourism hinges on a complex interplay of variables, which can be categorized into macroeconomic, microeconomic, and industry-specific dimensions. At the macroeconomic level, factors such as national GDP, currency exchange rates, and political stability play pivotal roles in shaping demand. For instance, the 2008 Global Financial Crisis demonstrated how economic downturns suppress discretionary spending, including high-cost tourism (Khandaker & Islam, 2017). Similarly, Le Goff and Moreau (2013) identified travel costs as the primary demand determinant, projecting that reducing ticket prices from 250,000 to 50,000 could expand the suborbital market from 600 to 128,500 customers by 2030. Regional disparities also emerge, with Americans and Chinese exhibiting the highest enthusiasm for space tourism, followed by Japanese, Australians, and Europeans, reflecting variations in disposable income and cultural perceptions of risk.

On the microeconomic front, individual financial capacity, health status, and psychographic motivations dominate decision-making. Psychological drivers such as the pursuit of self-actualization, scientific curiosity, and the allure of a unique experiential commodity further differentiate space tourists from conventional travelers (Vovchanska & Ivanova, 2022). Surveys indicate that 72% of prospective clients prioritize the symbolic value of spaceflight over logistical comforts, underscoring the industry's reliance on aspirational marketing.

However, space tourism introduces unique demand variables absent in terrestrial travel. Health risks associated with microgravity exposure (e.g., bone density loss, fluid redistribution) necessitate rigorous medical screenings, potentially excluding 30–40% of interested candidates (Musselman & Hampton, 2020). Operational constraints like limited spacecraft fleet sizes (2–6 passengers per launch), lengthy vehicle turnaround times (weeks to months), and regulatory coordination for airspace safety further cap scalability (Vovchanska & Ivanova, 2022). Crucially, the absence of orbital infrastructure beyond the ISS creates a "destination gap," though initiatives like SpaceX's Crew Dragon observation modules aim to mitigate this by 2030. These intertwined factors—economic, personal, and logistical—collectively define the market's contours, demanding tailored strategies for sustainable growth.

A comprehensive examination of the existing literature and relevant market reports has allowed for the identification of key variables that will significantly influence the demand factors in the space tourism industry. This study seeks to delve into a detailed analysis of these crucial elements and deliberate on the potential strategies for their cultivation and management within the context of space travel. By critically evaluating the interplay of these variables, this chapter endeavors to illuminate the factors that will shape and drive the demand for space tourism experiences, thereby contributing to a more nuanced understanding of the market dynamics and consumer behavior in this evolving sector. Furthermore, the investigation intends to explore the implications of these demand determinants on the development and sustainability of the space tourism industry, offering insights that can inform strategic decision-making and policy formulation to propel the growth and viability of this burgeoning field.

2 Literature Review

Market Demand

Mankiw analyzed various economic phenomena from both macro and micro perspectives. In his book "Principles of Economics", market demand is the sum of all people's demand for a good or service. From a micro perspective, it is the study of how households and firms make decisions that affect the market. The macro perspective, on the other hand, examines the impact of overall economic phenomena on the market. Especially, income, price, hobby, expectation, number of buyers, supply, etc., will affect the demand curve at the micro level. Market price controls, taxes, exchange rates, unexpected events, and policy changes can affect the market demand curve on a macro level.

All commodities (goods and services) that can be wanted are supposed to be specified by their physical characteristics in such a way that quantity and price are well-defined in the microeconomic theory of demand (Hildenbrand, 1994). Economics is not interested in the behavior of individual people; it is concerned with the behavior of groups of people. The study of individual needs is only one means of studying market demand. Macroeconomics is a subfield of economics that focuses on the behavior of the economy as a whole, including the markets, firms, customers, and governments. Macroeconomics studies trends in the economy, including inflation, price levels, economic growth rates, national income, GDP, and changes in unemployment. The two primary focuses of macroeconomic research are shorter-term business cycles and long-term economic growth.

Any whole market prediction is broken down into four steps (Barnett, 1988):

(1) Define the market first.
(2) Break down the overall industry demand into its constituent parts.
(3) Project the demand factors that more likely change for each category and forecast them.
(4) Perform sensitivity studies to identify the key hypotheses and assess the risks posed to the baseline projection.

Research on tourism demand over the past decade has undergone significant structural transformations and theoretical advancements. In terms of demand characteristics, traditional sightseeing tourism has progressively shifted toward experiential and themed travel, with independent travelers and family trips becoming dominant organizational forms, accompanied by heightened differentiation in consumption tiers (ITB China, 2024). Younger demographics (post-90 s and post-00 s cohorts) now constitute 40% of tourists, prioritizing "vitality-enhancing" immersive experiences over conventional attractions, driving a surge exceeding 200% in health-oriented retreats and nature-based tourism. Spatiotemporal analyses reveal differentiated seasonal fluctuations between coastal and inland regions, while the normalization of off-season travel and the emergence of high-demand clusters in southwestern China jointly validate the driving effects of economic development, transportation accessibility, and resource endowment.

The influencing mechanisms of tourism demand have evolved into multidimensional interactions. Economic factors demonstrate reconfigured weightings, where income levels and leisure time policies outweigh traditional Golden Week effects. Sociocultural dynamics highlight bifurcated demands: aging populations spur wellness tourism whereas youth demographics fuel adventure travel. The conceptual rise of transformative tourism (Kim et al., 2022) reflects value shifts toward "purpose-driven" journeys that simultaneously reshape individual perspectives and foster community development. Environmental quality and digital technologies emerge as critical variables: destinations with lower noise pollution gain prominence through "silence tourism," while VR/AR technologies enable blended virtual-physical consumption scenarios.

Methodological innovations leverage artificial intelligence and multisource data integration to enhance predictive accuracy. Traditional models like SARIMA and VAR continue to optimize cyclical pattern identification, while deep learning architectures (e.g., CNN-BiLSTM) significantly improve generalization capabilities in nonlinear demand forecasting through search engine data (SED) integration (Li et al., 2023). Hybrid modeling approaches, exemplified by Bayesian FAVAR models, reconcile quantitative predictions with qualitative insights by incorporating expert judgments and real-time data streams.

Emerging research prioritizes post-pandemic demand restructuring and techno-ethical challenges. Health security concerns solidify the normalization of self-guided tours and localized "micro-vacations," while redirected international demand accelerates rural educational tourism innovations. Sustainability frameworks now emphasize tourist carbon footprint accounting and destination carrying capacity management, with off-peak pricing strategies for aurora tourism in Nordic regions serving as representative case studies. Notably, despite enhancing personalized recommendation efficiency, big data and AI technologies provoke interdisciplinary ethical debates regarding data privacy breaches and algorithmic biases, necessitating balanced frameworks that harmonize technological applications with humanistic considerations.

Regional comparisons confirm strong positive correlations between economic development and demand intensity, yet lower-tier cities exhibit concentrated demand patterns. Policy formulation requires differentiated supply strategies: affluent regions should prioritize digital-cultural tourism innovations, while less-developed areas must capitalize on ecological and cultural assets. Additionally, establishing demand fluctuation early-warning systems and regional collaboration mechanisms will prove critical in mitigating overtourism pressures. Future research should investigate the long-tail effects of extreme events, metaverse-driven consumption reconfiguration, and cross-border demand synergies under global value chain restructuring, thereby providing forward-looking guidance for both theoretical and practical advancements.

Space Tourism

According to Chang (2017), the experience of weightlessness and the ability to observe the stars define space tourism, a branch of commercial space travel used for leisure and amusement. A commercial space sector has existed since 1970. The need for satellite-based services, such as television, GPS, and earth observation missions for search and rescue, is increasing globally each year (Chavagnac & Laporte-Weywada, 2009; Lappas, 2006). The space tourism sector has a lot of promise and seems to be a significant economic driver thanks to its capacity to reduce the cost of space access and attract new players to the space market. In recent decades, space travel has expanded significantly.

The space tourism market is divided into five categories by Laing and Crouch (2004): virtual, terrestrial, near-space, suborbital, and LEO/HEO. However, the discussion in this study is limited to the final two categories—SST and OST, 4 and 5 respectively. After more than 20 years of development, the main differences between these two categories are that (1) tourists in the former only travel in the suborbital, while tourists in the latter travel in suborbital, (2) tourists will travel to LEO in the latter, and (3) the cost per tourist-trip of the former will only be about one hundredth that of the latter.

Launch, spaceport, orbital destinations, training, and private astronaut mission recruitment and administration are some service categories catering to the space tourism business. Only the launch service provider's ability to carry clients to and from space is considered in This study. Suborbital Space (Blue Origin, Virgin Galactic) and Orbital Space (SpaceX, Roscosmos) are the two categories of launch service providers. The consumer will go in suborbital space to an altitude of more than 80 km before returning to Earth on a parabolic trajectory. A typical orbital space journey takes the passenger to 350 km or more altitude, then circles the planet until it docks with the space station or lands. The International Space Station (ISS) is where orbiting space travelers typically end up.

According to the Ganesha and Sinnoor (2021) analysis, most people trust and eagerly anticipate the moment we can travel to space. Few people also fear it and feel negatively about any potential harmful effects. Nonetheless, most people are enthusiastic about space travel. Twenty-two consumers paid up to an estimated $846 million in suborbital and orbital space tourism between 2001 and 2021 (20 years). 64% of the predicted clients and 67% of the projected sales were for 2021. All research proves that space tourism is an industry with great growth potential.

Space Tourism Demand

Space tourism necessitates a paradigm shift in tourism demand theory by challenging conventional assumptions about destination boundaries, rational decision-making, and sustainability metrics. Its transcendence of terrestrial geography—redefining orbital zones as destinations—and extreme risk profiles expose limitations in existing models that prioritize utilitarian consumer behavior. The tension between elite exclusivity and aspirational democratization reveals multidimensional value hierarchies, where tourists seek existential meaning and ontological security beyond functional satisfaction. Simultaneously, the interdependence of technological feasibility and psychological risk normalization processes invites theoretical synthesis between material constraints and perceptual constructs. Furthermore, the stateless nature of space environments demands an expanded conception of carrying capacity, integrating intergenerational equity into extraterrestrial resource governance. These challenges position space tourism as a catalyst for interdisciplinary convergence, bridging tourism studies with aerospace sociology, risk philosophy, and planetary ethics to develop adaptive frameworks that address both human-centric and post-anthropocentric dimensions of demand.

Numerous research has been conducted on tourism to understand better the various elements that drive the desire for travel. Seven categories—infrastructural, economic, political, environmental, customer-related, seasonality, vacation policies, and random- have influenced tourism demand studies (Liu et al., 2021). Song et al. (2023) thought the literature review method explains the variables used to determine tourism demand from unique perspectives (macro- and micro-level). The primary determinants of tourism demand at the macro level are GDP, exchange rate, and relative prices. At the micro level, factors like age (including generation and senior dummies), income, travel companions (including travel party size and children), past visits (first or repeat visit), price, and seasonality are the most significant person and travel-related features (including time trend, weekend, and holiday effects).

Space tourism is an emerging industry that faces unique challenges due to the lack of a fully developed industry chain and its current high cost, high price, and low availability. These factors have hindered the growth and development of space tourism as a viable sector. Unlike traditional tourism, the measurement of demand for space tourism cannot rely on conventional factors. One of the primary obstacles to the development of space tourism is the absence of a comprehensive industry chain. Unlike other forms of tourism, space tourism requires a complex network of infrastructure, technology, and services to support its operations. This includes spacecraft manufacturing, launch facilities, training centers, and specialized personnel. Currently, the industry is still in its infancy, with limited infrastructure and resources available. The lack of a complete industry chain poses a significant challenge to the growth of space tourism. Another significant barrier to the expansion of space tourism is its high cost, which limits its accessibility to only a select few. The development and operation of spacecraft, along with the necessary safety measures, require substantial investments. As a result, the price of space tourism experiences remains exorbitant, making it unaffordable for the majority of individuals. Furthermore, the limited availability of space tourism experiences further hampers its development. Currently, only a handful of companies offer space tourism services, and the number of available seats is extremely limited. Unlike traditional tourism, where demand can be measured based on factors such as destination popularity, accommodation bookings, and travel trends, space tourism presents unique challenges in demand measurement. The limited availability and high cost of space tourism experiences make it difficult to gauge the true demand for such services. Zhang and Wang (2022) review the current literature on space tourism and explore macro and micro perspectives in studying the economics of space tourism.

At the micro level, the subjects concerned new business potential, funding, and support for the space tourism sector. As opposed to prior government and taxpayer support, public–private partnerships (Beery, 2012; Collins, 2002), venture capital (Danilyan & Dzeban, 2019), and going public (Smith, 2000) have been proposed as new sources of investment. Particularly, the costs associated with choosing and building a spaceport site were outlined by Ingle (2011). Also, Goehlich (2005) and Komerath et al. (2007) simulated the cost of airline tickets and the revenue stream for space hotel companies, respectively. Due to the economies of scale effects, it is anticipated that space tourism will eventually facilitate all forms of space trips and become a crucial economic industry (Webber, 2013).

The macroeconomic effects of space tourism, including job creation, tax revenue generation, supply chain expansion, and related industry growth, have been extensively discussed (Collins & Autino, 2010; Friel, 2020; Webber, 2013). Cole (2015) asserted that space tourism would have the highest economic multipliers of any sector. Space tourism would indirectly aid the creation of space infrastructure and increase the scope of outer-erospace commercialization (Komerath et al., 2007). In addition, space tourism can benefit the economy of the tourist spots on Earth near the launch sites (Friel, 2020; Ingle, 2011; Spector, 2020). The impacts of space tourism on the current political-economic geography (Beery, 2012) and neoliberal capitalism paradigm (Valentine, 2012) would be substantial in addition to the industry's appealing economic generator roles, providing tourism economics with entirely new research areas.

Current research on demand comes from the following areas: product supply, customer motivation, customer experience, and customer adoption challenges. Regarding product supply, space tourism companies currently offer high-end luxury experiences or extreme adventure experiences. So, what motivates people to want to travel to space? A study conducted by the PEW Research Centre in 2018 provides some insight into customer motivations, which identified three main motivations for customers to buy space flights: to observe the Earth from space, to have an extraordinary experience (e.g., ground-breaking, exceptional, one-of-a-kind), and to be curious about the rest of the globe. Despite having the resources and desire to travel to space, clients still face many challenges, including risk, personal time commitments, no available orbital destinations, legal restrictions, health, language, etc.

So, how big is the market for space tourism? The fundamental barrier to customer acceptance that can be used to estimate market size is ticket price. For orbital customers, *space tourism market research* (Starzyk, 2022) has determined that an individual's net worth should not be more than 10% of the cost of a ticket to be regarded as a potential consumer. Tickets for suborbital consumers shouldn't cost more than 1.5% of a person's net worth. Starting at $450,000, suborbital space travel is a popular option. This implies that possibly viable clients would be people with an income of at least $30 million (approximately 200,000 individuals). Starting at $50 million, orbital space tourism passes are available. This implies that possibly prospective clients would be people with $500 million or more in assets (5304 individuals).

3 Methodology

Research Design

This study was proposed to be conducted through a literature review method using Pickering and Byrne (2014) research process, which was divided into five steps. (1) Determine the goals and objectives of the research, (2) then create the review criteria and process, (3) then conduct a search and refinement of the literature, then (4) extract and analyze the material, (5) and finally synthesize and present related findings. Firstly, a literature search and reading analysis was conducted to identify the factors influencing space tourism demand. This is followed by a close reading of the literature and a specific analysis of each influencing factor to explore how they will affect space tourism demand.

According to the research objective mentioned in the introduction section, a series of research procedures were set. First, search the SSCI database in Web of Science, and the term used to search This studys' titles, abstracts, and keywords was "space tourism." As can be seen from Fig. 1, research on space tourism began to take off gradually in 2010, after only one article was published in the SSCI in 2005.

Therefore, This study limits the literature search to 2010–2022, and three databases were selected (Science Direct, Web of Science, and Scopus) (Yang et al., 2017). The articles were selected in English by searching for "space tourism" in the title, abstract, and keywords, and the types of articles included academic journal articles and review articles.

Data Collection

201 papers containing space tourism were searched from 2010 to 2022, including 75 papers from Science Direct, 53 from Web of Science, and 73 from Scopus. All

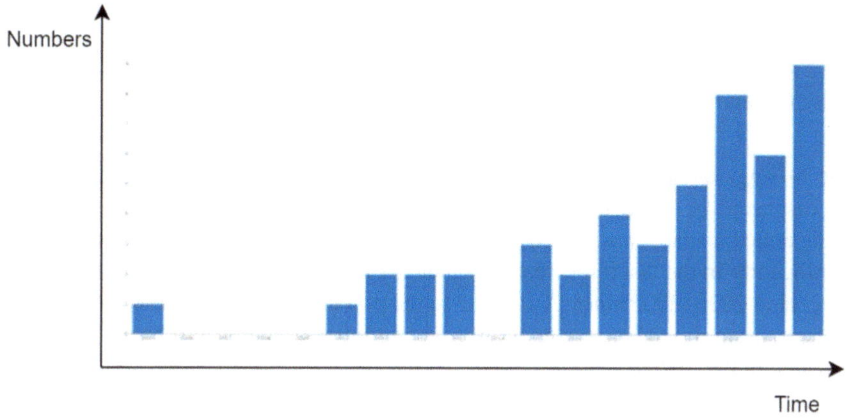

Fig. 1 Trend of space tourism articles published

literature was managed through the literature manager Endnote 20 software, and 128 literature remained after removing 73 duplicates. The authors carefully studied the titles, abstracts, findings, and conclusions of the remaining 128 publications to eliminate anything that did not meet the selection criteria. Even though they may also cover other forms and features of space travel, qualified publications must be based on the social sciences research paradigm and have at least one element specifically linked to demand issues for space tourism. Finally, the authors selected 21 studies that met the criteria.

Quantitative and qualitative analysis were used in this review's data analysis process. In detail, the quantitative studies were mainly descriptive, including (1) year of publication, (2) title of the journal, (3) type of journal, and (4) research methods used by the journal.

In terms of the qualitative analysis, a hybrid theme analysis was used to categorize and discuss study domains. First, the 6-step inductive theme creation procedure was modified for this survey, starting with the inductive subthemes (Braun & Clarke, 2006): Data familiarization, initial code creation, initial theme generation, initial theme review, initial theme confirmation, and report production are the first six steps. The chosen papers' abstracts, findings, and conclusions were read thoroughly and classified. The entirety of some intellectual pieces and brief thoughts were read and coded. The original sub-themes representing the specific research fields were then developed by combining all primary codes. The research was categorically summarized based on the requirements of space tourism in the following stage. All subthemes, or those specific research areas, were coded using this framework. Remembering that an article may include multiple research areas, and some sub-themes unrelated to the central theme or the research goals were eliminated.

4 Results

Descriptive Analysis

The findings of the quantitative descriptive analysis of space tourism publications are displayed in Table 1 Overall, there is a growing trend in the amount of study being done on space tourism. In particular, the number of publications has increased significantly in the last three years, with 38.10% (n = 8) of the number published in the period 2020–2022 since 2010. Due to the potential growth and continued use of space tourism, it is plausible to assume that additional studies will be undertaken in this area in the 2020s.

Regarding journal publication, space-focused journals (n = 12, 57.14%) published more space tourism research than. tourist industry journals (n = 7, 33.33%) over the past two decades. Acta Astronautica is the preeminent aerospace-social sciences interdisciplinary journal (n = 8, 38.10%). Comparatively, the distribution of similar publications in the category of travel journals is erratic. Some articles were published in law-focused publications, which are included in the category of others, in addition to space and tourism journals. Though space and tourism publications are the most important sources, space tourism studies may also be published in several interdisciplinary journals.

An analysis of the research methods found that most studies used qualitative and conceptual analysis (n = 16, 76.19%). Only two articles used qualitative and quantitative research methods, while three used quantitative research. However, none

Table 1 Descriptive analysis results

	2010–2014	2015–2019	2020–2022	Total
Total	**12**	**1**	**8**	**21**
Tourism journal	**2**		**5**	**7**
Tourism management	2			**2**
Tourism recreation research			1	**1**
International journal of tourism research			1	**1**
Journal of sustainable tourism			1	**1**
Journal of travel research			1	**1**
Tourism recreation research			1	**1**
Space	**9**	**1**	**2**	**12**
Acta astronautica	7		1	**8**
Space policy	1	1		**2**
Journal of space safety engineering			1	**1**
Journal of aeronautics, astronautics and aviation	1			**1**
Others	**1**		**1**	**2**

Table 2 Research methods analysis

Methods	2010–2014	2015–2019	2020–2022	Total
Qualitative	6		3	**9**
Quantitative			3	**3**
Mixed method	2			**2**
Conceptual	4	1	2	**7**

of these articles examined the demand for space tourism exclusively. All the results are shown in Table 2.

Thematic Analysis

After a descriptive analysis, a thematic analysis was carried out through a critical reading of the article. As there are not particularly many articles on just the space tourism demand variable, not all the selected literature examines the space tourism demand variable in isolation. It also involved many articles that explored only some of the factors. This study, from the macro- and micro-aspects, discusses the variables that will affect the space tourism demand. At the macro-level, it conducts policy, supply, and social media content. For the micro-level, it conducts risk, motivation, price, health, training, national, age, and gender.

Macro Variables

(1) Policy

Policy issues are a significant variable currently affecting demand for space tourism. These include policy imperfections, the division of responsibilities, flight altitude, and national airspace issues (Abou Yehia & Schrogl, 2010; Hobe, 2010; Masson-Zwaan & Freeland, 2010). Current regulations on space tourism are at the national level, whereas a commercial space trip involves subjects from several countries. For example, the country of production of the launcher, the operator, the launching country, and the landing country may apply four national subjects simultaneously. Abou Yehia and Schrogl (2010) research suggests that the generation of joint regulations organized by the European authorities would facilitate Europe's rapid capture of the space tourism market. At the same time, the issue of space legislation in cross-border countries is worth exploring (Masson-Zwaan & Freeland, 2010), while the definition of flight altitude for space tourism will also involve national airspace issues (von der Dunk, 2011). These relevant policy imperfections will affect the uncertainty of the legal status of tourists, the rights of stakeholders, etc. (Padhy & Padhy, 2021). When the rights of stakeholders are uncertain, the demand for tourism changes dramatically.

(2) Supply

The market demand for space tourism is intricately connected to the supply of available seats for travelers. The current carrying capacity of space travel is a crucial factor in determining the supply side of the industry. At present, the number of seats available for space tourism tourists is extremely limited, with only a few dozen seats being offered each year. This scarcity of supply poses significant challenges for the development of a robust and sustainable space tourism industry (Chang, 2015). With only a few dozen seats available each year, the competition for these seats is fierce, driving up the prices and making space tourism experiences exorbitantly expensive. This high cost acts as a deterrent for potential customers, limiting the accessibility of space tourism to only a select few who can afford it. The limited supply and high cost create a vicious cycle, as the high prices further restrict the demand, making it difficult for the industry to expand and attract more participants.

In order to overcome these challenges, it is crucial to increase the carrying capacity of space travel. This can be achieved through technological advancements in spacecraft design and manufacturing, as well as the development of more efficient and cost-effective launch facilities (Chang & Chern, 2013). By increasing the number of seats available for space tourism, the industry can attract a larger customer base and create a more sustainable market demand. Additionally, efforts should be made to reduce the cost of space tourism experiences, making them more accessible to a wider range of individuals.

In conclusion, the limited supply of seats for space tourism presents significant challenges for the formation of a complete industry chain and the accessibility of such experiences. Increasing the carrying capacity of space travel and reducing the cost of space tourism are crucial steps in overcoming these challenges and fostering the growth and development of the space tourism industry. By addressing these issues, we can make space tourism more accessible and sustainable in the future.

(3) Social Media Content

Promoting knowledge about space tourism on social media platforms has the potential to significantly influence people's perceptions and attitudes towards this emerging industry. The lack of knowledge and understanding about space tourism can often discourage individuals from considering or pursuing the opportunity to travel to space (Reddy et al., 2012). Therefore, it becomes crucial for space tourism suppliers to actively engage with scientists and experts in the field to disseminate accurate and relevant information through various media channels, including documentaries, news articles, and social media platforms.

Tasci et al. (2021) have shown that providing specific and scientific information about space travel generates a high response and impact on individuals' interest and willingness to participate in such experiences. By collaborating with scientists, space tourism suppliers can tap into their expertise and credibility to enhance the appeal and credibility of their offerings. Scientists can provide valuable insights into the technological advancements, safety measures, and potential benefits of space tourism, thereby addressing any concerns or misconceptions that potential travelers may have.

Social media platforms, with their wide reach and ability to engage with diverse audiences, offer an ideal avenue for distributing this knowledge. Suppliers can leverage social media channels to share informative content, such as videos, articles, and infographics, that highlight the scientific aspects of space travel. These platforms also provide an opportunity for interactive discussions and Q&A sessions, allowing potential travelers to directly engage with experts and clarify any doubts or queries they may have.

In addition to social media, documentaries and news articles can serve as powerful tools for educating the public about space tourism. Collaborating with filmmakers and journalists to produce and distribute informative and engaging content can help create a broader awareness and understanding of the industry. Documentaries can showcase the behind-the-scenes aspects of space travel, including spacecraft manufacturing, astronaut training, and the overall experience of being in space. News articles can cover the latest developments and breakthroughs in space tourism, further piquing the interest of potential travelers.

By actively promoting knowledge about space tourism through various media channels, suppliers can not only increase awareness and interest but also address any misconceptions or fears that may hinder individuals from considering space travel. This concerted effort to educate the public can ultimately contribute to the growth and sustainability of the space tourism industry by expanding the customer base and fostering a positive perception of this unique and transformative experience.

Micro Variables

(1) Motivation

Space tourism is an emerging market, and tourist motivations will largely influence tourist demand. Seeing the Earth from space is the motivation cited by most tourists (Frost & Frost, 2022; Reddy et al., 2012; Ziliotto, 2010), in addition to feeling weightless or zero gravity experiences and the unusual nature of the experiences are the most important reasons for traveling to space (Reddy et al., 2012). Olya and Han (2020) research found that the five motivations that people enter the space include information acquisition, gratification, adventure, service experience, and social. Wang et al. (2021) employed push and pull theory to analyze visitor motivation and concluded that pull elements included science, the grandeur of the universe, and space experiences. Push aspects included thrill, uniqueness, challenge, and escape.

(2) Price

Price is an essential factor in measuring demand, and specific measures of demand forecasts for space tourism are currently calculated through ticket prices, which are also a significant impediment to the civilianization of space tourism (Le Goff &

Moreau, 2013). Chang and Chern (2013) research found the current price threshold of OST is exceptionally high. Reducing costs through technological developments is the key to the current civilianization of space tourism. Improvements to the technical route of launch vehicle development could reduce the cost of access to orbit by a factor of 1000 within 15 years, allowing a new era of space flight (Ashford, 2019).

(3) Risk

Risk and safety factors are crucial to whether tourists want to go into space. Risk is the most significant reason why people are reluctant to undertake space tourism (Reddy et al., 2012), and the main categories in terms of people's specific perceived risks are human inadequacy to survive the launch and reentry phases of the journey, as well as personal inadequacy in terms of fitness and health to function in space. (Wang et al., 2021), therefore, reducing the perceived level of risk for travelers would increase their willingness to travel (Cates, 2021). On the one hand, it is possible to set up rescue measures to reduce the perceived risk, how to rescue people in the event of a crisis in space, and who should be rescued first, all of which need to be discussed and developed (Lyall, 2010). In addition to physical risks, financial risks are also a cause for concern. Space travel is expensive and carries with it the risk of being unable to travel in the short term after paying for it, in addition to the possibility of losing a lot of extra costs (Olya & Han, 2020).

(4) Healthy

The health concerns and effects are mainly physical and psychological (Reddy et al., 2012). On the physical side, first of all, space travel has high physical requirements for visitors, who first need to pass a medical examination (Chang & Chern, 2013); in addition to requiring a qualified physical condition, the microgravity environment and space radiation during the launch and landing can cause physical damage (Kluge et al., 2013). These are all huge challenges for private spacecraft passengers. Regarding psychological health, the enclosed space environment, spatial constraints, circadian rhythms, and sleep effects can cause anxiety and psychological stress to travelers (Oluwafemi et al., 2021; Olya & Han, 2020). These health concerns and negative impacts could create significant travel restrictions, which could affect the willingness of space passengers to travel and hence the demand for travel.

(5) Training

Another limitation of access to space tourism is that space tourism requires training and is not as straightforward to undertake as other travel can be. There are two main aspects of training for travel: the length of training and the strict training (Reddy et al., 2012). Pre-trip training for space travel takes a long time; most passengers are willing to spend 2–4 weeks training (Reddy et al., 2012), but the training takes 9–18 months (LEO/HEO). The other is the strict nature of the training and the knowledge base required, which consists of general space training, specialized space training, and preflight space training. Each stage requires a great deal of learning and physical training, which is also quite challenging for the passengers (Chang & Chern, 2013).

(6) Demographics of the Passengers

Demographic characteristics impact demand in three main variables: gender, age, and nationality. Reddy et al. (2012) study found that 74% of women were not interested in space travel. In the younger group, women were concerned about the risks that would negatively affect their enthusiasm for traveling. In contrast, in the male group, the risk factor did not affect their confidence for space travel. Also, in the female sample, the main variables influencing motivation were the uniqueness, challenge, and the beauty of the universe (Wang et al., 2021). Secondly, in terms of age, the over 35 s are less interested in space tourism, while the younger age groups are very enthusiastic about it (Reddy et al., 2012). Different nationalities also have different attitudes towards space tourism, which may be due to different cultural backgrounds. Safety, environment, and self-indulgence can discourage Europeans from space tourism, but this is not evident for Americans and Asians. Adventure is the main factor affecting Germans but not significantly for the Japanese, with pioneer divers being the most crucial reason for the Japanese (Le Goff & Moreau, 2013).

5 Conclusion and Future Research

Conclusion

This study begins with a quantitative analysis summarizing the year of publication and journal type of research articles on space tourism demand. In terms of year of publication, the number of articles related to space tourism demand research has been increasing day by day since 2020, indicating that space tourism, as an emerging market, has risen in recent years and has excellent research potential.

Secondly, more than half of the articles were published in journals in the space field, and only seven articles were published in tourism journals, indicating that research on space tourism demand is still under development. Particularly in the field of tourism, research on space tourism is not popular. Still, as the topic of space has increased in popularity in recent years, researchers have begun to focus more and more on this area.

An analysis of the research methods of the selected articles reveals that most of the current studies are still at the qualitative and conceptual stages. In particular, the articles before 2020 primarily discuss phenomena, related policies, and current issues. Very little deals with specific analysis on the issue of research needs. In recent years, articles have gradually started using quantitative research methods to explore space tourists' motivations, behaviour and so on. However, there is still no specific research on the variables that influence the demand for space tourism.

The nine main variables influencing demand were extracted from carefully reading the literature in the thematic analysis. A classification by macro and micro revealed that the macro variables included policy, supply, and social media content. Micro variables included motivation, price, risk, health, training, and demographics of the passengers. The main demographic variables are income, gender, age, and nationality of the passengers.

Theoretical and Practical Contributions

Exploring demand-related variables has helped deepen the understanding of consumer behavior in niche markets such as space tourism. This enriches academic work in fields such as tourism research, consumer psychology and behavioral economics. Understanding demand variables supports economic analysis and helps assess market potential and financial viability. Understanding Demand Variables supports comprehensive market feasibility studies that provide valuable data for entrepreneurs and investors. Insights into demand-related variables inspire innovative business models and marketing strategies. Understanding demand variables enriches academic engagement in aerospace engineering and technology research by fostering discussion of scientific advances and technological innovations. Understanding the demand variable facilitates a critical assessment of the ethical and environmental impacts associated with space tourism. By understanding the variables associated with the demand for space tourism, scholars, industry stakeholders, and policymakers can collectively contribute to the responsible and sustainable development of this emerging industry while advancing the cross-fertilization of knowledge from different disciplines.

Future Research

Simultaneously, the current state of research predominantly occupies a conceptual and qualitative realm, featuring limited quantitative analyses concerning space tourism demand. Future research endeavors are poised to pivot towards a more quantitative orientation, focusing on gauging the impact of existing variables on space tourism demand. This shift is anticipated to offer a more granular understanding of the intricate interplay between various determinants and their influence on the burgeoning sphere of space tourism. Furthermore, there exists a notable gap in the literature, as evidenced by the scarcity of dedicated articles probing the variables that steer space tourism demand. Consequently, future research initiatives hold promise in delving deeper into unearthing and comprehensively understanding the specific determinants that underpin the demand for space tourism.

Moreover, an intriguing avenue for prospective investigation lies in the cultural disparities between Eastern and Western societies, potentially yielding distinct attitudes and viewpoints towards space tourism. Recognizing these cultural differentials prompts the necessity for further empirical inquiry to unravel potential divergences and convergences in perceptions and predispositions regarding space travel. By undertaking comprehensive research endeavors that encompass cultural variations, scholars can enrich the current discourse, fostering a more nuanced comprehension of the multifaceted landscape of space tourism demand within a global context. Therefore, future research endeavors should embrace this cultural dichotomy, elucidating the complexities that stem from diverse cultural outlooks and their impact on the burgeoning domain of space tourism.

References

Abou Yehia, J., & Schrogl, K.-U. (2010). European regulation for private human spaceflight in the context of space traffic management. *Acta Astronautica, 66*(11), 1618–1624. https://doi.org/10.1016/j.actaastro.2009.10.024

Ashford, D. (2019). Corporate groupthink: The main obstacle to an affordable Lunar Base. *Space Policy, 50*, Article 101344. https://doi.org/10.1016/j.spacepol.2019.101344

Barnett, F. W. (1988). Four steps to forecast total market demand. *Harvard Business Review, 66*(4), 28–38.

Beery, J. (2012). State, capital and spaceships: A terrestrial geography of space tourism. *Geoforum, 43*(1), 25–34. https://doi.org/10.1016/j.geoforum.2011.07.013

Braun, V., & Clarke, V. (2006). Using thematic analysis in psychology. *Qualitative research in psychology, 3*(2), 77–101. https://www.tandfonline.com/doi/abs/10.1191/1478088706QP063OA

Cates, G. (2021). The in-space rescue capability gap [风险]. *Journal of Space Safety Engineering, 8*(3), 202–210. https://doi.org/10.1016/j.jsse.2021.07.007

Chang, Y.-W. (2015). The first decade of commercial space tourism. *Acta Astronautica, 108*, 79–91. https://doi.org/10.1016/j.actaastro.2014.12.004

Chang, Y. W. (2017). A preliminary examination of the relationship between consumer attitude towards space travel and the development of innovative space tourism technology. *Current Issues in Tourism, 20*(14), 1431–1453. https://doi.org/10.1080/13683500.2015.1005580

Chang, Y. W., & Chern, J. S. (2013). From spirit of St. Louis to SpaceShipTwo: A study on the challenge of future space tourism development [Article]. *Journal of Aeronautics, Astronautics and Aviation, 45*(2), 135–154. https://doi.org/10.6125/12-1120-715

Chavagnac, C., & Laporte-Weywada, H. (2009). *The suborbital space tourism project of eads astrium*. 45th AIAA/ASME/SAE/ASEE Joint Propulsion Conference & Exhibit.

Cole, S. (2015). Space tourism: Prospects, positioning, and planning. *Journal of Tourism Futures, 1*(2), 131–140.

Collins, P. (2002). Meeting the needs of the new millennium: Passenger space travel and world economic growth. *Space Policy, 18*(3), 183–197. https://doi.org/10.1016/S0265-9646(02)00018-8

Collins, P., & Autino, A. (2010). What the growth of a space tourism industry could contribute to employment, economic growth, environmental protection, education, culture and world peace. *Acta Astronautica, 66*(11), 1553–1562. https://doi.org/10.1016/j.actaastro.2009.09.012

Danilyan, O., & Dzeban, A. (2019). Space law at 21st century: The security issues. *Philosophy and Cosmology-Filosofiya I Kosmologiya, 22*, 6–15. https://doi.org/10.29202/phil-cosm/22/1

Friel, M. (2020). Tourism as a driver in the space economy: New products for intrepid travellers. *Current Issues in Tourism, 23*(13), 1581–1586. https://doi.org/10.1080/13683500.2019.1628189

Frost, J., & Frost, W. (2022). Exploring prosocial and environmental motivations of frontier tourists: Implications for sustainable space tourism. *Journal of Sustainable Tourism, 30*(9), 2254–2270. https://doi.org/10.1080/09669582.2021.1897131

Ganesha, K. S., & Sinnoor, G. B. (2021). Sentimental analysis of space tourism: Evidence from Twitter. *New Space-the Journal of Space Entrepreneurship and Innovation, 9*(3), 148–150. https://doi.org/10.1089/space.2020.0052

Goehlich, R. A. (2005). A Ticket pricing strategy for an oligopolistic space tourism market. *Space Policy, 21*(4), 293–306. https://doi.org/10.1016/j.spacepol.2005.08.007

Le Goff, T., & Moreau, A. (2013). Astrium suborbital spaceplane project: Demand analysis of suborbital space tourism. *Acta Astronautica, 92*(2), 144–149. https://doi.org/10.1016/j.actaastro.2013.03.025

Hildenbrand, W. (1994). Market Demand. In *Market demand: Theory and empirical evidence* (pp. 30–71). Princeton University Press. https://doi.org/10.1515/9781400863716.30

Hobe, S. (2010). The legal regime for private space tourism activities—An overview. *Acta Astronautica, 66*(11), 1593–1596. https://doi.org/10.1016/j.actaastro.2009.08.019

Hurley, B. (2022). *Space tourism market analysis*. https://newspaceeconomy.ca/2022/02/08/space-tourism-market-analysis/

Ingle, M. (2011). The copernican shift in space tourism and its implications for tourism in the Great Karoo. *South African Geographical Journal, 93*(1), 104–118, Article Pii 937651673. https://doi.org/10.1080/03736245.2011.561543

Khandaker, S., & Islam, S. Z. (2017). International tourism demand and macroeconomic factors. *International Journal of Economics and Financial Issues, 7*, 389–393.

Kluge, G., Stern, C., Trammer, M., Chaudhuri, I., Tuschy, P., & Gerzer, R. (2013). Commercial suborbital space tourism-proposal on passenger's medical selection. *Acta Astronautica, 92*(2), 187–192. https://doi.org/10.1016/j.actaastro.2012.08.005

Komerath, N., Nally, J., & Zilin Tang, E. (2007). Policy model for space economy infrastructure. *Acta Astronautica, 61*(11), 1066–1075. https://doi.org/10.1016/j.actaastro.2006.12.033

Laing, J., & Crouch, G. (2004). *Vacationing in space: Tourism seeks "new skies"*. CABI International. https://doi.org/10.1079/9780851998633.0011

Lappas, V. (2006). Space tourism. In *Tourism business frontiers* (pp. 157–167). Routledge.

Liu, H., Liu, W., & Wang, Y. (2021). A study on the influencing factors of tourism demand from mainland China to Hong Kong. *Journal of Hospitality & Tourism Research (Washington, D.C.), 45*(1), 171–191. https://doi.org/10.1177/1096348020944435

Lyall, F. (2010). Who is an astronaut? The inadequacy of current international law. *Acta Astronautica, 66*(11), 1613–1617. https://doi.org/10.1016/j.actaastro.2009.11.017

Masson-Zwaan, T., & Freeland, S. (2010). Between heaven and earth: The legal challenges of human space travel. *Acta Astronautica, 66*(11), 1597–1607. https://doi.org/10.1016/j.actaastro.2009.12.015

Musselman, B. T., & Hampton, S. (2020). Factors influencing the emergence of suborbital space tourism. *International Journal of Aviation, Aeronautics, and Aerospace, 7*(2), 9. https://doi.org/10.58940/2374-6793.1493

Oluwafemi, F. A., Abdelbaki, R., Lai, J. C. Y., Mora-Almanza, J. G., & Afolayan, E. M. (2021). A review of astronaut mental health in manned missions: Potential interventions for cognitive and mental health challenges. *Life Sciences in Space Research, 28*, 26–31. https://doi.org/10.1016/j.lssr.2020.12.002

Olya, H. G. T., & Han, H. (2020). Antecedents of space traveler behavioral intention. *Journal of Travel Research, 59*(3), 528–544. https://doi.org/10.1177/0047287519841714

Padhy, A. K., & Padhy, A. K. (2021). Legal conundrums of space tourism [法律]. *Acta Astronautica, 184*, 269–273. https://doi.org/10.1016/j.actaastro.2021.04.024

Pickering, C., & Byrne, J. (2014). The benefits of publishing systematic quantitative literature reviews for PhD candidates and other early-career researchers. *Higher Education Research & Development, 33*(3), 534–548. https://doi.org/10.1080/07294360.2013.841651

Reddy, M. V., Nica, M., & Wilkes, K. (2012). Space tourism: Research recommendations for the future of the industry and perspectives of potential participants. *Tourism Management, 33*(5), 1093–1102. https://doi.org/10.1016/j.tourman.2011.11.026

Smith, V. L. (2000). Space Tourism: The 21st Century “Frontier.” *Tourism Recreation Research, 25*(3), 5–15. https://doi.org/10.1080/02508281.2000.11014920

Song, H., Qiu, R. T. R., & Park, J. (2023). Progress in tourism demand research: Theory and empirics. *Tourism management (1982), 94*, 104655. https://doi.org/10.1016/j.tourman.2022.104655

Spector, S. (2020). Delineating acceptable risk in the space tourism industry [Article]. *Tourism Recreation Research, 45*(4), 500–510. https://doi.org/10.1080/02508281.2020.1747798

Starzyk, S. S. B. J. (2022). *Space tourism market study.*

Tasci, A. D. A., Fyall, A. D., & Fu, X. (2021). Social representations of space travel: Modeling the antecedents and outcomes [Article]. *International Journal of Tourism Research, 23*(4), 611–635. https://doi.org/10.1002/jtr.2430

Valentine, D. (2012). Exit strategy: Profit, cosmology, and the future of humans in space. *Anthropological Quarterly*, 1045–1067.

von der Dunk, F. G. (2011). Space tourism, private spaceflight and the law: Key aspects. *Space Policy, 27*(3), 146–152. https://doi.org/10.1016/j.spacepol.2011.04.015

Vovchanska, O., & Ivanova, L. (2022). Marketing analysis of diversification of the factors determining the market conditions of the space tourism market. *Turystyka i Rozwój Regionalny* (18), 129–138. https://doi.org/10.22630/TIRR.2022.18.25

Wang, L., Stepchenkova, S., & Kirilenko, A. P. (2021). Will the present younger adults become future orbital space tourists? *Tourism Recreation Research, 46*(1), 109–123. https://doi.org/10.1080/02508281.2020.1762342

Webber, D. (2013). Space tourism: Its history, future and importance. *Acta Astronautica, 92*(2), 138–143.

Yang, E. C. L., Khoo-Lattimore, C., & Arcodia, C. (2017). A systematic literature review of risk and gender research in tourism. *Tourism Management, 58*, 89–100. https://doi.org/10.1016/j.tourman.2016.10.011

Zhang, Y., & Wang, L. (2022). Progress in space tourism studies: A systematic literature review [Article]. *Tourism Recreation Research, 47*(4), 372–383. https://doi.org/10.1080/02508281.2020.1857522

Ziliotto, V. (2010). Relevance of the futron/zogby survey conclusions to the current space tourism industry. *Acta Astronautica, 66*(11), 1547–1552. https://doi.org/10.1016/j.actaastro.2009.08.027

Chapter 3
How Is Nonlinearity in Space Tourism Demand Modeled?

1 Introduction

By definition, space tourism is space travel by passengers for recreation (Chang, 2020). As space tourism is a very new field, the literature conducted in this field is much less than in other tourism fields (Danov, 2020; Zhang & Wang, 2022). Since the first human travel into outer space in the mid-1960s, various passengers have traveled to outer space aboard rockets or space shuttles. However, most passengers have a work-related purpose, with essential tasks of scientific research or piloting the vehicle. Only a smaller number of these passengers go into space for recreational purposes (usually, they are millionaires), that is, for space tourism (Cohen & Spector, 2020; Masson-Zwaan & Freeland, 2010). Thus, space tourism is relatively new, and the twenty-first century is the century of space travel, which is related to the maturation of the market for space tourism and the rising demand for space travel services (Cohen & Spector, 2020). Estimating the demand in the newer areas is beneficial to estimate the size of the market that is not yet mature and make trend judgments. In addition, space tourism is the fast-growing tourism field that could help humans immigrate to Mars in 2035. Therefore, space tourism is expected to have the potential for more mature market mechanisms and more passenger volumes by then. Therefore, there are theoretical implications for estimating and modeling demand before the market is mature, as it not only lays the methodological foundation for modeling by potential future researchers, but also facilitates taking full advantage of the current maturity conditions for a stage-by-stage theoretical summary, and in doing so, prepares future researchers for theoretical expansion.

Much of the current research addressing the assessment of tourism demand is based on linear models or other more straightforward perspectives (Wen et al., 2019), as is the case in space tourism. (Carbonell et al., 2009). Therefore, a simple linear review is beneficial in allowing practitioners to quickly predict and judge the likely future trends and direction of the industry. However, nonlinear space tourism forecasting or assessment is necessary. This is not only because nonlinear assessment

K. Peng et al., *Space Tourism Supply Chain*, Contributions to Management Science,
https://doi.org/10.1007/978-981-95-1080-1_3

can be closer to the actual demand situation compared to linear prognosis, but also because the application of nonlinearity can make the definition of demand types more valuable (Chen et al., 2017).

The application of nonlinear models in space tourism demand analysis addresses the inherent complexity of this emerging market. Unlike traditional linear approaches, nonlinear frameworks are uniquely suited to capture the multidimensional dynamics shaping demand. Price elasticity in space tourism exhibits asymmetric characteristics, where high-income consumer segments may demonstrate price insensitivity, while specific pricing thresholds in broader markets could drive nonlinear demand surges. Technological advancements such as reusable launch systems and regulatory shifts in space governance often create discontinuous demand patterns, requiring modeling tools capable of detecting phase transitions and threshold behaviors. Nonlinear methodologies enable the identification of synergistic interactions between variables like safety perceptions and disposable income, which may amplify or suppress demand through multiplier effects. These models also adapt to nonstationary market conditions, particularly during the nascent stage of industry development where demand trajectories may follow sigmoidal growth patterns rather than linear projections. The dynamic learning capacity inherent in advanced nonlinear systems proves critical for tracking real-time market evolution, a vital feature given the fluid business models and rapid technological iterations in space tourism. While acknowledging computational and interpretational challenges, the strategic value of nonlinear modeling lies in its ability to uncover latent market tipping points and inform adaptive pricing frameworks, making it an essential analytical paradigm for this high-potential sector.

Therefore, this study aims to explore the application of nonlinear modeling in space tourism demand and plans to reach this chapter purpose through the following three steps.

(1) Review the historical literature on nonlinear modeling of space tourism demand.
(2) Summarize the historical literature's main points, nonlinear application scenarios, and research purposes.
(3) To summarize and analyze the information covered above and point out the current research directions with research value.

2 Literature Review

Abitzsch (1996) argued that the demand for space tourism can be revealed through computer simulations. In other words, the demand for space travel is revealed employing computer simulations that exhibit the characteristics of a disposition curve on a medium time scale. The authors analyze the demand of four scenarios over the next 50 years by assuming three scenarios and adding a realistic technological scenario as a reference. Vertical Take-Off Vertical Landing (VTOVL) is considered to be a powerful technique to stimulate the demand for space tourism based on the data. Escher (1995) suggest that VTOVL and HTOHL are considered suitable for acceptance into the "mixed mode" of space transportation. This is because of the advantages of using this mode over conventional modes in terms of development and operating costs. However, both VTOVL and VTOVL + HTOHL were determined by the computer to be in line with the S-shaped demand curve. That is, although the rollout of the technology will increase the demand for space tourism for a specific period, a slowdown in the growth rate is sure to come.

Crouch (2001) builds on the above and finds that as space tourism costs improve, the demand for space tourism rises in a curvilinear fashion. More specifically, this study generates three curves: a low probability, a medium probability, and a high probability curve. In either curve, the market demand is significantly stimulated once the single ticket price is below $100,000. When the single-visit cost of space tourism exceeds $100,000, the weakening effect of the price increase on demand is much weaker. One possibility for this phenomenon is that both Abitzsch (1996) and (Collins et al., 1996) reveal that less than 20% of respondents are willing to spend a year's worth of income on space travel. A single trip costing $100,000 exceeds many household incomes and is a significant expense for these respondents to ignore.

Corporation (2002) found that for 450 respondents, the length of time needed for orbital space travel was quite elastic. The relationship between demand and price for space tourism at mid-orbit flight altitudes is not directly rigid and linear. When the price of a single trip to space orbit drops to $5 million and $1 million, the market demand rises significantly, and travelers' willingness to travel increases significantly. It is worth noting that the price of space orbit travel is expected to continue to decline over time from its current level of $20 million. In other words, the demand for space from travelers who can see orbiting vehicles in our mid-lifetimes will be firmly and powerfully stimulated for a short to medium period. In other words, this rise in demand will not increase gradually over time but will be concentrated at some point.

Crouch et al. (2009) showed that categorical data premises (e.g., the country of the operator launching the rocket/spacecraft) were also revealed to affect respondent willingness significantly. For example, in the belief of high-altitude travel, Australia's spaceships are considered to have the most positive impact on the spending intentions of potential customers compared to other countries. Meanwhile, Russia was identified as the least favored country of departure for space travel among various countries. This phenomenon is also reflected in zero gravity space travel, where potential consumers remain more hostile or skeptical of Russian space travel services. Interestingly, the effect of weightlessness on potential consumers is also not continuous. When the duration of weightlessness is less than 10 min (3min/ 4min/ 6min), the experiment results show that it does not significantly impact consumers. However, when the duration of weightlessness reached 10 min, the period of weightlessness proved to be an attractive point for potential consumers in suborbital space travel. For the independent variable of education level and the premise of consumption intention as the dependent variable, the image is characterized by a U-shaped curve. Those with low and high levels of education will have a greater interest in high-altitude flight tourism. Respondents in the middle of the spectrum show very little interest. In other words, the future promotion of space travel could be considered for those with higher or lower education levels. This is because those who have earned a bachelor's degree have a limited interest in space travel. In a survey of consumer intentions for suborbital space travel, the authors found no significant differences in motivation between men and women. In zero-gravity flight travel, the physical requirements for potential consumers were revealed not to affect consumer intentions significantly. This differed from ground travel, where physical limitations are an essential reason travelers abandon the destination or route (Molokáč et al., 2022).

Another point is that an essential purpose of zero-gravity travel is space tourism, and the type (owned/ shared) and size (big/ small) of the windows used for tourism could not significantly influence the respondents. In overtourism, oversharing core tourism resources can be fatal to the traveler's sense of experience (Deng & Ma, 2014; Parsaee et al., 2021). In contrast, the view of space tourism as one of the most critical trips in space tourism actually, in the eyes of potential consumers, does not mind sharing with others, which is sufficient to show a difference between terrestrial tourism and space tourism in this dimension. Consumers are more concerned about the launch training aspect of parachute training and launch training. This is because there is insufficient evidence that parachute training has a significant impact on the perception of consumption, which may be related to the fact that consumers willing to participate in space tourism are generally in better physical and psychological condition, which facilitates consumers to pass the basic physical and mental tests of space tourism (Cater, 2019; Scott Jr, 2000). Parachute training does not present much of a physical barrier or psychological barrier issue for this group. However, for launch training, the data from the respondents showed that they would mind extensive training. This is because simple training is revealed to benefit consumers' perceptions. On the contrary, extensive training is detrimental to consumers' mental perceptions.

In Launius (2019) study, the public's willingness to travel into space remained moderately high before the collapse of the former Soviet Union. When the Soviet Union collapsed, the public's desire to travel in space declined significantly and severely. That is, public attitudes toward space travel are not invariably linearly correlated. Public attitudes toward space travel were tied to the macro environment, such as the political environment of the time. In the early twenty-first century, Russia began offering tourist attractions for missions to the International Space Station to raise funds for its struggling space flight program. Eight (7 tourists) were conducted until 2009. Each passenger spent between $20 million and $40 million per one-way trip to space (Cohen & Spector, 2020). But then, no new passengers ever came to the door demanding space travel services.

In contrast, space travel in the United States has been much more successful, with space travel companies represented by SpaceX, Blue Origin, and others having made corresponding contributions to the commercial space travel industry (Chang, 2015; Mehran et al., 2023). Thus, the enthusiasm of some eligible space tourism tourists for space travel may be influenced not only by various objective or subjective conditions of individuals but also by the conditions of the countries from which the vehicles depart. The social effectiveness of space tourism commercial insurance is nonlinear. The initial launch of space travel insurance with few users will have a small and unstable effect on society. Only when the market and technology mature a larger sample will have the opportunity to rapidly increase the social effectiveness of space tourism commercial insurance (Bensoussan, 2010).

For the variables influencing the avoidance intention of space tourism, Olya and Han (2020) argue that it can be explained by the push-pull-mooring (PPM) framework. PPM is rooted in migration theory, which explains why people move from one place to another (Bansal et al., 2005). PPM is used as a holistic framework to stimulate individuals' shifting behavior or switching behavior in different disciplines, e.g., transportation (Wang et al., 2020) and aviation (Jung et al., 2017). The original migration theory only addressed the direct effects of push and pull on human behavior. In This study, however, the authors set the mooring as a moderator. It is assumed to positively moderate the effect of the push and pull factors on the dependent variable. One of the more seminal points of This study is that it is approached from the perspective of avoidance intention, rather than from the perspective of positive intention. Interestingly, the mooring variable not only directly affects avoidance intention, but actively modifies the effect of pull factor on avoidance intention. That is, there will be an interaction term in the theoretical model. The positive impact of this interaction term will replace the linear relationship in This study as an essential innovation. A similar effect does not hold for the push factor. That is, the mooring variable does not modulate the effect of push factor on avoidance intention. This result is consistent with the finding of Jung et al. (2017), that is, the impact of the mooring factor on the push factor is insignificant. In other words, one of the two interaction terms in This study has a significant effect on space tourism avoidance intention, and the impact of the other one is not statistically significant. This result shows that in the direct impact (linear effect), all three factors positively and significantly impacted space tourism avoidance intention. Still, in the nonlinear domain, this effect was only partially supported.

About extended model of goal-directed behaviour (EMGB), there are several reasons why EMGB is affected by the nonlinearity of the space travel experience. Such a research hypothesis is derived from Kim et al. (2023). Multi-group analysis (MGA) was applied to the above study. In other words, the amount of space travel experience was used as an extra-SEM variable to see if the relationship in different groups should change across groups. The final results show that the MGA results are negative and that the PLS results for the 16 paths do not differ enough in statistical willingness. That is, the amount of space tourism experience does not significantly affect the relationships of the variables in the hypothesized model. Interestingly, the results from the fsQCA run are not identical to the PLS-SEM.

3 Methodology

The process of literature selection is delineated into four sequential steps, strategically designed to streamline the identification and curation of pertinent resources for researchers in the field. Initially, papers pertaining to space tourism are sourced from prominent databases such as ProQuest, Google Scholar, and Scopus, ensuring a comprehensive pool of literature. To refine this selection, a crucial distinction is made between quantitative and non-quantitative research papers, as nonlinear modeling is exclusive to quantitative studies. Consequently, all conceptual papers are systematically excluded from consideration, given their non-quantitative nature. Similarly, qualitative papers are isolated, as they do not typically incorporate nonlinear modeling within their design. This meticulous pruning effectively eliminates both conceptual and qualitative papers from the database, ensuring a repository of literature aligned with the study's quantitative orientation.

Subsequently, a rigorous assessment is conducted to ascertain the utilization of nonlinear modeling within the remaining papers. Through a thorough examination of these papers, those lacking references to nonlinear modeling are methodically eradicated from the database. This discerning procedure ensures the preservation of only those materials that align with the specific quantitative scope of the study, maintaining precision and relevancy in the literature compilation.

Consequent to this meticulous filtering process, the culled literature is methodically organized into a comprehensive literature matrix. This matrix serves as a strategic tool, enabling researchers to traverse the developmental trajectory of the literature while unraveling its inherent internal logic. Each entry within the literature matrix encapsulates vital information and salient content derived from the selected literature, furnishing an invaluable resource for subsequent analyses and scholarly discourse within the realm of space tourism demand. The culmination of this meticulous process results in the comprehensive documentation and preservation of crucial literature in the forthcoming section, serving as an indispensable foundation for informed research and scholarly deliberations.

4 Results

Therefore, we have concluded the necessity of nonlinear models in space tourism research from the following aspects.

The demand for space tourism exhibits complex nonlinear dynamics influenced by multifaceted socioeconomic and experiential factors. While decreasing costs generally stimulate market growth, analyses reveal distinct patterns across pricing thresholds and consumer segments. Historical data indicates that space tourism demand follows an S-curve trajectory under optimal pricing scenarios, where initial cost reductions yield minimal market response until reaching critical price points (Yanguang, 2012). Specifically, when ticket prices fall below $100,000, demand accelerates exponentially due to increased accessibility for high-net-worth individuals and aspirational upper-middle-class consumers. This threshold represents a psychological tipping point where space travel transitions from being perceived as an exclusive luxury to an attainable once-in-a-lifetime experience.

The income-demand relationship demonstrates nonlinear characteristics influenced by disposable income distribution. Less than 20% of potential travelers consider allocating a full year's income for space tourism, creating a natural demand ceiling that varies across economic demographics. This financial constraint interacts with experiential factors—particularly weightlessness duration—in unexpected ways. Experimental studies reveal a minimum experiential threshold where suborbital flights offering less than 10 min of weightlessness fail to generate significant consumer satisfaction, while flights exceeding this duration show disproportionate increases in perceived value (Kim et al., 2024). This suggests that experiential quality operates through a step-function mechanism rather than linear progression.

Educational attainment introduces another layer of complexity through U-shaped consumption patterns. Both highly educated consumers (with greater scientific curiosity and risk tolerance) and those with basic education (driven by status-seeking motivations) demonstrate elevated interest compared to moderately educated middle segments (Witek & Kuźniar, 2020). This bifurcation reflects differing value perceptions, where advanced education correlates with intrinsic motivation for space exploration, while basic education associates with extrinsic social signaling benefits.

The interaction effects between variables generate nonlinear patterns that defy simple additive models (Hainmueller et al., 2019). For instance, the combination of moderate pricing (200,000 range) with extended weightlessness duration (12–15 min) generates multiplicative demand effects exceeding individual factor contributions. Conversely, high-income consumers with postgraduate education show price inelasticity below \$250,000 but extreme sensitivity to experiential compromises, illustrating how demographic intersections modify conventional economic assumptions.

Market expansion faces dual constraints from technological scalability and psychosocial barriers (Adner et al., 2016). While reusable launch vehicles continue reducing marginal costs, consumer anxiety about spaceflight safety creates a psychological demand barrier that follows inverse logistic patterns—initial safety improvements yield diminishing fear reduction returns until surpassing a critical reliability threshold. This phenomenon interacts with cultural factors, as collectivist societies demonstrate steeper safety-perception curves compared to individualistic cultures.

The emerging market structure suggests three distinct growth phases: (1) Elite accessibility (\$500 k + tickets) with linear demand growth, (2) Mass-premium phase (\$100 k-\$500 k) characterized by logistic adoption curves, and (3) Potential mass-market expansion (<\$100 k) requiring breakthrough innovations in both vehicle economics and social normalization of space travel. Current projections indicate the market will remain in Phase 2 through 2035, constrained by training time requirements and infrastructure limitations rather than pure pricing factors.

Demand forecasting for space tourism may require non-linear modelling to cope with multiple complex mechanisms: a step increase in demand when prices drop to a specific threshol, a bimodal distribution of willingness to consume due to differences in educational backgrounds, synergistic amplification of experience duration and pricing, and dynamic constraints bounded by income levels. These factors are intertwined to form a non-linear response network, which requires forecasting tools to break through the limitations of linear assumptions and capture market laws through flexible modelling to support decision-making.

5 Conclusions

The emergence of space tourism has led to a growing body of literature focused on understanding the factors that influence its pricing and potential inflection points (Abitzsch, 1996; Salkeld & Beichel, 1977). Previous research has suggested that price inflection points are significantly associated with the behavior of space tourists. To better comprehend this phenomenon, the Theory of Planned Behavior (TPB) may offer a more comprehensive framework for understanding space tourism behavior, going beyond simple research-based behavioral control. Indeed, the forecast of price inflection points in space tourism is believed to carry important implications for the industry's development.

Moreover, the application of the push-pull model in space tourism has predominantly focused on personal factors, as noted by the researcher (Jung et al., 2017). However, there is a growing recognition of the need for studies conducted at a social level, considering factors such as organizational reputation. Utilizing Hierarchical Linear Modeling (HLM) could provide valuable insights into individual motivational factors, particularly in relation to the push and pull factors associated with space tourism.

Furthermore, China's active participation in the development of space tourism offers an important opportunity for the industry, potentially serving as a catalyst for an inflection point. China's well-established and mature space industry chain positions it as a key player in the future of space tourism. This could significantly impact the dynamics of space tourism and play a pivotal role in ushering in an inflection point.

In conclusion, the literature on space tourism pricing and inflection points is expanding, with scholars investigating the application of behavioral models such as the TPB and push-pull model to better understand the complexities of space tourism behavior. As the industry continues to evolve, it is important to consider not only individual factors but also broader societal and industry-related influences, such as China's involvement, to anticipate and capitalize on potential inflection points in space tourism.

References

Abitzsch, S. (1996). Prospects of space tourism. *DGLR BERICHT*, 99–112.

Adner, R., Ruiz-Aliseda, F., & Zemsky, P. (2016). Specialist versus generalist positioning: Demand heterogeneity, technology scalability and endogenous market segmentation. *Strategy Science, 1*(3), 184–206.

Bansal, H. S., Taylor, S. F., & James, Y. S. (2005). "Migrating" to new service providers: Toward a unifying framework of consumers' switching behaviors. *Journal of the Academy of Marketing Science, 33*(1), 96–115. https://doi.org/10.1177/0092070304267928

Bensoussan, D. (2010). Space tourism risks: A space insurance perspective. *Acta Astronautica, 66*(11–12), 1633–1638. https://doi.org/10.1016/j.actaastro.2010.01.009

Carbonell, P., Rodríguez-Escudero, A. I., & Pujari, D. (2009). Customer involvement in new service development: An examination of antecedents and outcomes. *Journal of Product Innovation Management, 26*(5), 536–550. https://doi.org/10.1111/j.1540-5885.2009.00679.x

Cater, C. (2019). *History of space tourism*. Emerald Publishing Limited.

Chang, Y.-W. (2015). The first decade of commercial space tourism. *Acta Astronautica, 108*, 79–91. https://doi.org/10.1016/j.actaastro.2014.12.004

Chang, Y.-W. (2020). From aviation tourism to suborbital space tourism: A study on passenger screening and business opportunities. *Acta Astronautica, 177*, 410–420. https://doi.org/10.1016/j.actaastro.2020.07.020

Chen, W., Sun, J., Gao, S., Cheng, J.-J., Wang, J., & Todo, Y. (2017). Using a single dendritic neuron to forecast tourist arrivals to Japan. *IEICE Transactions on Information and Systems, E100.D*, 190–202. https://doi.org/10.1587/transinf.2016EDP7152

Cohen, E., & Spector, S. (2020). Space tourism-past to future: A perspective article. *Tourism Review, 75*(1), 136–139. https://doi.org/10.1108/TR-03-2019-0083

Collins, P., Stockmans, R., & Maita, M. (1996). Demand for space tourism in America and Japan, and its implications for future space activities. *Advances in the Astronautical Sciences, 91*, 601–610.

Corporation, F. (2002). *Space tourism market study: Orbital space travel & destinations with suborbital space travel*. Futron.

Crouch, G. I. (2001). The market for space tourism: Early indications. *Journal of Travel Research, 40*(2), 213–219. https://doi.org/10.1177/004728750104000212

Crouch, G. I., Devinney, T. M., Louviere, J. J., & Islam, T. (2009). Modelling consumer choice behaviour in space tourism. *Tourism Management, 30*(3), 441–454. https://doi.org/10.1016/j.tourman.2008.07.003

Danov, D. G. (2020). A review of space tourism services: supply and demand challenges. *Journal of Tourism Leisure and Hospitality, 2*(1), 29–35. https://dergipark.org.tr/en/pub/toleho/issue/53785/739977

Deng, T., & Ma, M. (2014). Resource curse in tourism economies? An investigation of China's world cultural and natural heritage sites. *Asia Pacific Journal of Tourism Research, 19*(7), 809–822. https://doi.org/10.1080/10941665.2013.806943

Escher, W. (1995). Rocket-Based Combined-Cycle (RBCC) powered spaceliner class vehicles can advantageously employ Vertical Takeoff and Landing (VTOL). In *International Aerospace Planes and Hypersonics Technologies*. https://doi.org/10.2514/6.1995-6145

Hainmueller, J., Mummolo, J., & Xu, Y. (2019). How much should we trust estimates from multiplicative interaction models? Simple tools to improve empirical practice. *Political Analysis, 27*(2), 163–192. https://doi.org/10.1017/pan.2018.46

Jung, J., Han, H., & Oh, M. (2017). Travelers' switching behavior in the airline industry from the perspective of the push-pull-mooring framework. *Tourism Management, 59*, 139–153.

Kim, M. J., Hall, C. M., Kwon, O., Hwang, K., & Kim, J. S. (2024). Orbital and sub-orbital space tourism: Motivation, constraint and artificial intelligence. *Tourism Review, 79*(2), 392–407. https://doi.org/10.1108/TR-01-2023-0017

Kim, M. J., Hall, C. M., Kwon, O., Sohn, K., & Kim, M. (2023). The effects of individual and organizational interventions on space tourism: Applying EMGB and fsQCA. *Current Issues in Tourism*, 1–5.

Launius, R. D. (2019). Human aspirations to expand into space. In *Space tourism* (Vol. 25, pp. 15–50). Emerald Publishing Limited. https://doi.org/10.1108/S1571-504320190000025002

Masson-Zwaan, T., & Freeland, S. (2010). Between heaven and earth: The legal challenges of human space travel. *Acta Astronautica, 66*(11), 1597–1607. https://doi.org/10.1016/j.actaastro.2009.12.015

Mehran, J., Olya, H., & Han, H. (2023). Psychology of space tourism marketing, technology, and sustainable development: From a literature review to an integrative framework. *Psychology & Marketing, 40*(6), 1130–1151. https://doi.org/10.1002/mar.21795

Molokáč, M., Hlaváčová, J., Tometzová, D., & Liptáková, E. (2022). The preference analysis for hikers' choice of hiking trail. *Sustainability, 14*(11), 6795. https://www.mdpi.com/2071-1050/14/11/6795

Olya, H. G. T., & Han, H. (2020). Antecedents of space traveler behavioral intention. *Journal of Travel Research, 59*(3), 528–544. https://doi.org/10.1177/0047287519841714

Parsaee, M., Demers, C. M. H., Potvin, A., Hébert, M., & Lalonde, J.-F. (2021). Window view access in architecture: Spatial visualization and probability evaluations based on human vision fields and Biophilia. *Buildings, 11*(12), 627. https://www.mdpi.com/2075-5309/11/12/627

Salkeld, R. J., & Beichel, R. (1977). Mixed-mode rocket vehicles for international space transportation systems. *Acta Astronautica, 4*(1–2), 213–227.

Scott, R. W., Jr. (2000). Policy/Legal framework for space tourism regulation. *J. Space l., 28*, 1.

Toivonen, A. (2021). Sustainable space tourism. *Channel View Publications*. https://doi.org/10.1080/09669582.2020.1783276

Wang, S., Wang, J., & Yang, F. (2020). From willingness to action: Do push-pull-mooring factors matter for shifting to green transportation? *Transportation Research Part d: Transport and Environment, 79*, Article 102242. https://doi.org/10.1016/j.trd.2020.102242

Wen, L., Liu, C., & Song, H. (2019). Forecasting tourism demand using search query data: A hybrid modelling approach. *Tourism Economics, 25*(3), 309–329. https://doi.org/10.1177/1354816618768317

Witek, L., & Kuźniar, W. (2020). Green purchase behavior: The effectiveness of sociodemographic variables for explaining green purchases in emerging market. *Sustainability, 13*(1), 209. https://doi.org/10.3390/su13010209

Yan-Guang, C. (2012). On the urbanization curves: Types, stages, and research methods.

Zhang, Y. Z., & Wang, L. R. (2022). Progress in space tourism studies: A systematic literature review. *Tourism Recreation Research, 47*(4), 372–383. https://doi.org/10.1080/02508281.2020.1857522

Chapter 4
How Are Intertemporal Dynamics in Space Tourism Demand Accounted For?

1 Introduction

The study of tourism economics has always been based on demand analysis. See, for instance, Li et al. (2005) for a review of the methods used to analyze tourism demand. Since the 1990s, demand modeling studies have shifted from static regression models to more sophisticated dynamic specifications. Dynamic models aim to avoid potential problems such as spurious regression, poor prediction, and structural instability (Song et al., 2019), and consider important factors such as repeat visits, habit persistence, word-of-mouth recommendations, or reputation (Morley, 2009). These models support the idea that previous visitors influence the current tourism demand. This study refers to this intertemporal relationship as a persistence or reputation effect, which can be attributed to various factors.

The most common way to incorporate dynamics into demand models is to include lagged demand as an explanatory variable. However, this may not be sufficient to explain the dynamics of travel demand (Morley, 2009). Even simply including lagged demand assumes persistence or reputation effects. Tourism demand responds to changes in influencing factors simultaneously and to the time lag caused by the persistence of habits. For example, lagged tourist numbers can be used to capture structural dependencies and habit persistence; lagged GDP can explain the effects of long-term economic convergence in source markets; and lagged values of search queries can be used to model search behavior by forecasting travel demand.

Existing studies have shown that integrating autoregressive variables and distributional lags among influencing factors can be used to determine intertemporal dynamic patterns of tourism demand. Information criteria, such as the Akaike Information Criterion (AIC), Bayesian Information Criterion (BIC), Schwarz Information Criterion (SIC), Hannan-Quinn Criterion (HQC), or Last Prediction Error, which provide researchers with a degree of deterministic time series (dependent or independent) is a long sequence of memories. It also gives them confidence that the integration relationships between variables in the requirements model are robust.

K. Peng et al., *Space Tourism Supply Chain*, Contributions to Management Science,
https://doi.org/10.1007/978-981-95-1080-1_4

The dynamic concept of intertemporal selection can be explained as follows. In addition to external environmental events, an individual's decision to travel abroad is influenced by past travel experience and future travel expectations. Additionally, the size of the current cumulative travel budget results from travel decisions made in recent years. That is, the current optimal choice will change based on the choices made in other years. However, the most popular tourism development theory of the Tourism Area Life Cycle (TALC) shows that delayed demand's effect on actual demand is not constant. Still, it increases as the employment level approaches the target carrying capacity varies (Albaladejo et al., 2016). According to TALC theory, visitors' growth rate is faster in the initial stage. However, this process slows down as it approaches carrying capacity.

However, intertemporal dynamic analysis also has disadvantages. Another normative issue related to the intertemporal character of demand models is the stationarity of the data used in model estimation. A time series is said to be stationary if its statistical properties, such as mean, variance, and autocorrelation, are time-invariant. Time series analysis using non-stationary data can produce spurious regressions and draw flawed conclusions. Individuals choose which goods or services to purchase based on their most excellent satisfaction or benefit from consumption. Given budget and time constraints, most early research focused on the concurrent trade-offs between time and money. In reality, however, many decisions involve trade-offs between current and future payoffs (i.e., intertemporal choices). Passengers can travel abroad during the current year rather than deferring it to the next year or the year after. In addition to anticipated future journeys, actual journeys are influenced by previous travel experiences. No travel in the last year may increase a traveler's propensity to travel that year due to accumulated travel budgets and desires. Optimal intertemporal travel choices are made based on the degree to which travelers perceive and value utility when approaching past, present, and future time horizons.

Many scientific data, various meta-analyses, and review papers tease out theories and methods of forecasting tourism demand (Crouch, 1995; Peng et al., 2014; Wu et al., 2017). However, few studies currently use econometrics to explore market demand forecasting for space tourism. Hence, to fill the highly desirable research gap, this study may show how knowledge and theories are advanced under intertemporal choices in the existing tourism and hospitality literature. A perspective to further expand space tourism's economic development and market demand.

The rest of this chapter is structured as follows. The second part sorts out the application of intertemporal dynamics, an econometric method, in tourism. The third part introduces the literature analysis method and the reasons for the research selection. The fourth part is the research results and discussion of intertemporal dynamics, an econometric method, in tourism. The last section concludes.

2 Literature Review

Space Tourism Market Demand Analysis

Marketing research is critical to commercial space tourism as these trips are designed to understand consumer behavior towards such services (Moreau et al., 2001). Since space tourism is technically feasible, many variables need to be analyzed from the perspective of entrepreneurs and potential consumers. For example, the relationship between customer demand and pricing and the most critical set of characteristics of a space tourism experience will determine the amount a customer pays. Attention is turning to the travel price and the experience's safety aspects. Spaceflight is still dangerous, and all visitors must acknowledge that they face relatively high risks to their health. The stress on the body during take-off landing, the effects of weightlessness, and the fact that the entire journey takes place in a relatively small space are characteristics that can affect the health of even the most persistent human. The physical and mental stress testing is more detailed in the case of orbital flight. In contrast, suborbital travel requires only a basic set of checks.

We are witnessing the birth of an emerging commercial space tourism industry. While the road ahead seems highly uncertain, a wide range of economic, technological, political, legal, environmental, financial, and commercial issues ultimately determine the pace and direction the industry needs to take. Industry experts are acutely aware that at the forefront of these uncertainties is how consumers perceive and respond to the competitive choices on offer. For example, many in the industry expect prices to fall, thus making the sector more mainstream. In this context, a valid, reliable, and accurate estimate of market demand is critical to attracting the interest, participation, and cooperation of many different businesses, organizations, and individuals, particularly in the capital markets. While venture capitalists such as Boston Harbor have supported space tourism ventures, attracting venture capitalists and private equity firms has proved more difficult. These investors are nervous because commercial human spaceflight is risky. Still, because the company competes with traditional software, internet, biotech, and medical device startups, these ventures are less challenging and potentially more rewarding. Validating the market for public space travel in general and true tourism will be an essential step.

Data source: Compiled by the author.

Table 1 outlines the research methodology applied to space tourism market research in the last two decades (2001–2022). Regarding research orientation, 38 studies are conceptual, three are qualitative, ten are quantitative, and three are mixed-method. As shown in Table 1, most of the studies are conceptual because of data collection challenges.

Table 1 Literature on space tourism research methods

Research method	Number of publications	Authors
Conceptual	38	Beery (2012), Chang (2015), Forganni (2017), and Margolis (2020)
Quantitative	10	
Survey (close-ended)	7	Olya and Han (2022) and Tasci et al. (2021)
Econometrics modeling of pricing strategy (secondary data)	1	Goehlich (2005)
Econometrics modeling of expectations (secondary data)	1	Duan (2020)
Econometrics modeling of transportation for long-duration space tourism (secondary data)	1	Savino et al. (2013)
Qualitative	3	
Interviews	3	Cater (2010), Frost and Frost (2021), and Soleimani et al. (2018)
Mixed-method	3	
Survey & interviews	3	Jacobs et al. (2019), Reddy et al. (2012), and Toivonen (2020)

Application of Intertemporal Dynamics in the Field of Tourism

Within the traditional typology of forecasting strategies, there are four main approaches to tourism forecasting (van Doorn, 1984). Survey methods (e.g., time series analysis, historical analogy, causal methods, projection scenarios, and morphological analysis), speculative methods (e.g., Delphi, group consensus, brainstorming, or individual expert opinion) and social construction methods (e.g., subjective probability forecasting, Bayesian statistics, pattern recognition or expected scenarios) fall into these categories. Based on a review of empirical studies, Witt (1995) identifies three types of forecasting models: causal models (econometric and spatial models), time series models, and various qualitative methods. Recently, it has been widely accepted that there are four approaches to forecasting tourism demand (Song & Li, 2008). The three groups of quantitative methods are time series models, econometric models, and artificial intelligence-based models (Goh & Law, 2011; Peng et al., 2014). Judgment methods that can be used for both qualitative and quantitative forecasting belong to the fourth group (Lin & Song, 2014). Regression analysis methods dominated the tourism forecasting and modeling literature until the 1990s. In recent years, econometric forecasting models have become known for their greater accuracy as they integrate recent developments in econometric methodology (Song & Li, 2008). Other quantitative methods, such as gravity models, artificial neural networks (ANNs), and univariate time series models also play a crucial role in tourism demand forecasting. However, there is still disagreement about which models provide the most reliable forecasts under various conditions. Each method has advantages in solving a particular problem, but none has proven inherently superior.

The basic econometric models are mainly divided into static econometric models (i.e., regression models, gravity models, Static Almost Ideal Demand System (AIDS)) and dynamic econometric models (i.e., Vector Autoregressive (VAR), Time-Varying Parameter (TVP), Error Correction Model (ECM), Cointegration and Autoregressive Distributed Lag Model). Although static econometric models are more powerful in analyzing and explaining elasticities, they are less effective in predicting tourism demand because their estimates undermine long-term relationships and short-term variability (Smeral et al., 1992). For this reason, in recent years, more and more scholars have begun to pay attention to dynamic econometric models and try to use them to predict tourism demand, such as Vector Autoregressive (Gunter & Önder, 2015), Time-Varying Parameter (Gunter & Önder, 2015; Song et al., 2011), Error Correction Model(Gunter & Önder, 2015; Vanegas Sr & Vanegas, 2013) and Autoregressive Distributed Lagged Model (Gunter & Önder, 2015; Tukamushaba et al., 2013) and TVP structural time series model (Gunter & Önder, 2015; Song et al., 2011). The introduction of causality has improved the predictive power of these forecasting models, enabling them to capture continuous changes in customer tastes and preferences (Peng et al., 2014). These models also explain the intertemporal relationships between tourism demand and its many influences. In particular, distributed lag models consider current android values of factors influencingurrent tourism demand (Guizzardi & Stacchini, 2015; Wan & Song, 2018). To this end, this study collates research on intertemporal dynamics in tourism (shown in Table 2.)

3 Methodology

In this study, the literature analysis was conducted by utilizing the Web of Science database to search for articles containing the keywords "intertemporal dynamics" or "intertemporal choices" in combination with "tourism," "space tourism," or "hospitality" within the full text. This methodological approach aimed to identify relevant scholarly works that specifically addressed the interplay of intertemporal dynamics within the realm of tourism and hospitality, with a particular focus on potential implications for the burgeoning field of space tourism. The search yielded a total of eight journal articles published from 2000 to 2023, which were initially subjected to rigorous screening criteria to eliminate irrelevant studies, resulting in the identification of eight pertinent articles for in-depth analysis.

Given the relatively limited number of feasible articles identified through this process, the typical use of citation analysis software such as CiteSpace may not be practical in this context. Therefore, the research strategy pivoted towards a comprehensive content analysis of the selected literature to elucidate the application of intertemporal dynamics within these studies. This approach aimed to delve into the contextual conditions under which the dynamic econometric analysis method of

Table 2 Literature on intertemporal dynamics in tourism

Literature/ Authors	Study context	Key findings
Gabriel et al. (2008)	Advertising	The study shows that in the short term, increased marketing expenditure raises foreigners' demand for tourism, leading to higher relative prices for tourism services, making tourism production more attractive and stimulating capital investment. Over time, the capital stock increases, and tourism production expands, leading to lower tourism prices. In the long run, increased marketing activity leads to higher tourism productivity, higher capital stocks, lower relative prices for tourism services, and reduced net foreign assets
Hsu et al. (2013)	International tourist air travel	The results support the argument that adverse external events such as emerging flu pandemics or economic crises harm international travel demand. Furthermore, the results suggest that travelers tend to wait until their savings budgets have built up enough to cover the cost of travel
Bunghez (2015)	Space Tourism Market	Space tourism is currently only open to the very wealthy. The most promising part of space tourism open to the general public is suborbital flights led by Virgin Galactic. Based on the research in this study, it can be observed that there is a high demand for such activities from the public. As more operators of such services enter and consumer traffic increases, prices will fall, and the space tourism sector will become one of the leading tourism activities that generate profits
Albaladejo et al. (2016)	Reputation effect	The results show that the reputation effect is not constant in either estimate, supporting the idea that tourist congestion affects the number of visitors to Spain
Wong et al. (2019)	Gambling	The findings illustrate the combined effect of individual-level and organizational-level characteristics on the direct and moderating effects of gambling decisions. In particular, brand equity moderated the relationship between demographic and event-related characteristics and gambling decisions

(continued)

Table 2 (continued)

Literature/ Authors	Study context	Key findings
Abrate et al. (2019)	Hotel revenues	This study aims to analyze the impact of dynamic price variability on revenue maximization. Although dynamic pricing is a common practice in the tourism industry, this is the first comprehensive study of the relationship between price variability and revenue maximization. Within the framework provided by the literature on intertemporal price discrimination, price equity, inventory control, and organizational culture, this study proposes and applies a new hedonic revenue model with a sample of 21,687 observations
Karl et al. (2021)	Tourist's impulse buying behavior	This study critically assesses how tourists' impulse buying behavior has been conceptualized and tested, discusses advances in the field, and sets out a research agenda to investigate the relationship between impulse buying and tourism decisions
Abdou et al. (2021)	Tourism demand forecasting techniques	This study reviews 145 key papers from 1979 to 2020 to better understand how tourism demand forecasting techniques have changed. The three forecasting model types are econometric, time series, and artificial intelligence (AI). The main conclusion from the historical comparison of forecasting methods is that forecasting models have become more diverse, merged, and improved forecasting accuracy

Data source Compiled by the author

intertemporal dynamics was employed, the underlying rationale for its utilization, the specific methodological techniques deployed, as well as the key research findings gleaned from these studies. Furthermore, the investigation sought to identify potential avenues for expansion and extension in future research, aiming to provide a robust foundation for advancing the incorporation of intertemporal dynamics within the domain of space tourism research.

The selected articles were meticulously reviewed to extract key insights into the application of intertemporal dynamics within the context of tourism and hospitality, particularly with relevance to the evolving landscape of space tourism. This involved a systematic examination of the theoretical frameworks, methodological approaches, empirical findings, and theoretical contributions within each study, allowing for a comprehensive understanding of how intertemporal dynamics have been integrated into the scholarly discourse within this domain.

The analysis also sought to elucidate the underpinning motivations and justifications for employing intertemporal dynamics as an analytical tool within the examined literature. This entailed a critical assessment of the methodological choices, including the specific econometric models utilized, the temporal dimensions considered, and the analytical techniques employed to capture and analyze intertemporal dynamics within the context of tourism and hospitality. By scrutinizing the methodological intricacies and empirical outcomes of these studies, the research aimed to discern the substantive contributions and implications of intertemporal dynamics for enhancing our understanding of the interplay between temporal dynamics and decision-making processes within the realm of space tourism.

Furthermore, the exploration of potential extensions and amplifications for future research perspectives played a pivotal role in the analysis. This involved identifying gaps, limitations, and unexplored avenues within the existing literature, offering valuable insights into the untapped potential for further integrated study of intertemporal dynamics within the evolving landscape of space tourism research. By delineating these future research directions, the study aimed to provide a roadmap for advancing the integration of intertemporal dynamics as a foundational analytical paradigm within the domain of space tourism research, thus contributing to the enrichment and expansion of scholarly discourse in this burgeoning field.

4 Analysis Results and Discussions

Intertemporal Dynamics Approach Applied

The above articles conclude that the research backgrounds include the impact of tourism advertising activities on the macroeconomy, international tourism demand preferences, non-constant reputation effects in dynamic tourism demand models, forecasting of gambling demand by exhibition tourists, and the impact of price changes on income. Maximizing impact, impulse buying behavior in travel and so on.

The Reasons to Use the Intertemporal Dynamics Approach

In the first study, the authors found that many countries currently invest a lot of resources in advertising campaigns to promote their resources to increase their market share. Consumers see advertising messages everywhere they watch television, read magazines, listen to the radio, go online, or walk down the street. However, past research has explored less about spending on tourism demand marketing, and less research on tourism demand has included advertising. Whereas changes in demand may translate into changes in market prices, which in turn affect the economic decisions of other economic agents, coupled with the fact that the impact of these marketing activities may spill over into other sectors, it is more reasonable to analyze the economic impact of marketing activities comprehensively through an intertemporal framework. In the second study, individuals choose goods or services based on their consumption of the good or service that generates the most excellent satisfaction or utility and on considerations of budget time; most previous studies have involved decisions about the trade-off between time and money. In reality, however, many decisions involve trade-offs between current and future benefits (i.e., intertemporal choices). Travelers often decide whether to travel abroad in the present or postpone the decision until the next year or years in the future. In addition, current travel behavior is influenced by past travel experiences, i.e., the possibility that not having traveled in the previous year may increase the traveler's propensity and desire to travel in the past year. Based on this, the intertemporal dynamics model can better capture the full range of past, present, and future processes based on the extent to which travelers perceive and weigh utility over a close time horizon. In the third study, which focuses on a review of space tourism, intertemporal dynamics are mentioned in the measurement methodology to predict the economic and social potential of space tourism behavior. In the fourth study, based on the Tourism Area Life Cycle (TALC) theory, the most popular theory on the evolution of tourism, it is shown that the effect of lagged demand on current demand is not constant but varies in carrying capacity as the level of occupancy approaches the destination. This process slows down as it approaches that carrying capacity. The ratio of tourists to carrying capacity is the cause of tourism congestion.

How to Use Intertemporal Dynamics for Analysis

The second article uses a binary choice model to investigate the travel choice behavior of international air travel. The Cobb-Douglas functional form is applied in this study as it is widely used to describe the utility function when variables are correlated and correlated. Let us denote the unit of time by year in the study, with t representing a specific year. Current travel choices include travel choices and non-travel choices, where travel choices are those where the tourist travels abroad for a holiday in the current year. In addition, instead of traveling in the current year, a travel decision can be deferred to future years, where it is denoted as a non-travel choice. The travel budget in this study is defined as the total amount of money that can be allocated to holiday travel in a year. Individuals with different socio-economic characteristics have different travel budgets. Retirees may have a lower income from work but may have a surplus in their travel budget due to savings and pensions.

On the other hand, working individuals may have high incomes but little surplus to spend on travel due to heavy taxes and consumption of other goods. This study focuses on the intertemporal choice of international air travel. Individual socio-demographic variables have not been included to simplify the study. However, the impact of an individual's socioeconomic characteristics and an individual's age on travel demand behavior was determined by using a travel budget in a pragmatic model. This study included travel budgets and holiday days to capture an individual's ability to travel abroad. Indicators representing the individual's ability to travel in money and time in years t, m_t, *and* n_t are then used.

Findings

In the first study, we can find that increasing marketing activities can increase the relative price of tourism services in the short term, stimulate capital investment, and lead to a gradual increase in tourism production. With the frequent occurrence of economic deficits, dynamic adjustments are closely related to real exchange rate depreciation. In the long run, the economy's final net foreign asset stock is low, the capital stock is high, the production of tourism services increases accordingly, and the real exchange rate appreciates. It can be seen that the impact of marketing on domestic agent consumption and the resulting welfare effects are not significant. In short, analyzing the economic structure before engaging in relevant marketing activities is especially necessary. The second study found that past travel experience has a greater impact on future travel decisions. In addition, tourists often need to wait until their savings budget has accumulated enough to cover travel expenses before traveling. Taking Taiwanese tourists as an example in this study, it can be seen that high-income tourists in Taiwan travel internationally almost every year.

The fourth study analyzed the panel data of the number of tourists in 17 autonomous regions of Spain from 2000–2013. It used two different GMM estimators, the difference GMM and the system GMM, respectively, for the number of domestic tourists and the number of international tourists is analyzed. In both cases, the econometric model includes traditional economic factors, such as income and relative prices, and a quadratic function of lagged demand. The sixth study builds on an extensive theoretical literature in four areas: intertemporal price discrimination, price fairness, inventory control, and organizational culture. Since few studies overcome data scarcity, this study measures whether and to what extent intertemporal price changes favor final income. First, dynamic price variability has a positive impact on revenue despite the presence of strategic customers and perceptions of inequity (Aviv & Pazgal, 2008). Second, the study proposes a hedonic income model that exploits dynamic price variability to address endogeneity. Third, it is helpful for hotel practitioners to have a clearer understanding of OTA's revenue maximization strategy using the agency model. It can be seen that making correct price decisions is particularly important for individual hotels' transaction performance based on the agency model. The eighth study summarizes the development pattern of the four forecasting methods after assessing the core research on tourism demand modeling and forecasting. As tourism demand forecasting continues to attract the attention of scholars, forecasting methods have grown exponentially. Despite emerging trends and new approaches to demand forecasting, time series, intertemporal dynamics, econometrics, artificial intelligence, and hybrid models still dominate tourism demand forecasting. Baseline tests of the predictive performance of hybrid models outperform single standard models. Combined forecasting is increasingly common, and numerous studies have shown that it improves accuracy. Nonlinear combination for single-component forecasting models is a new and superior approach to traditional linear combination methods in terms of combination techniques.

Prospects for Future Research

In the first study, the current tourism advertising marketing only considers stakeholders such as multiple departments. Still, it has not been based on tourists' perspective because tourists' habits may also cause tourism demand to have an impact on marketing activities. In the second study, the study analyzed the data collected during the 911 attacks and the outbreak of SARS and found that adverse external events will lead to a decline in outbound travel demand. In contrast, positive external events may increase air travel demand. Thus, future research could investigate how other environmental variables, such as infrastructure and tourism competition, affect tourism demand. In addition, the utility function in this study considers travel costs, travel time, budget, and vacation time. However, travel is also affected by personal characteristics, income, education level, and even national GDP. Future research can also consider the above variables within range. In the fourth study, since this study mainly focuses on international tourism, it ignores the situation of domestic tourists and residents. Compared with international tourism, the impact of domestic tourism on different archipelagos in Spain may be different. In addition, future research can also explore the demand analysis of different tourism motivations from the perspective of tourists. In the fifth study, given that most commercial events are only held during the day, event participants only have time for leisure activities such as gambling in the evening.

Future research may, therefore, extend this chapter by exploring event-specific attributes (e.g., venue location and accessibility, service quality and event type, etc.) that influence patrons' gambling behavior. In the seventh study, future research can consider the intertemporal decision-making model as a general framework to further study the impulse purchase of tourism products. Researchers can further evaluate the temporal extension of impulsive behavior and measure how it is related to cognitive processing system interaction. For example, the COVID-19 pandemic has led to the interruption and suspension of tourism activities. During this period, people's desire to travel will primarily cause consumers' impulse buying behavior. Future research can also explore how expectations affect the travel experience need. In the eighth study, as the number of innovative models for tourism demand forecasting grows, hybrid models incorporating new forecasting models will become possible to improve forecasting accuracy. Since the study aims to examine the evolution of the forecasting process, the determinants of tourism demand, the variables affecting the elasticity of demand, and the estimated impact of intervention events on various scenarios can also be taken into account.

5 Conclusions

In general, the rise of the "space tourism movement" will have a significant cultural impact and will open people's eyes uniquely to the twenty-first century. Under this influence, space development will acquire an economic dimension and mark the true beginning of the "space age," where people will arrive at this frontier as passengers, customers, employees, operators, and managers, and not just as government employees. In this way, roles will swap, and consumers and providers of private space services will become the majority.

The beginning of suborbital tourist flights will open the way for a revolution in how we see and manage society. It is hard to imagine where this first step will take us. Just as the advent of the automobile 100 years ago revolutionized the world and gave rise to unexpected industries, we can predict a similar transformation in human activity in space. Fully understanding that these innovations will transform our society, inspired by previous examples, potential investors in this industry will generate huge profits in the long run, even if the initial cost of participating in this race seems unreasonable. The research shows that potential customers demonstrate the demand for space travel and the financial availability of this activity. The results encourage investors in space tourism, as prices are still high and the requirements to conduct this type of business are not entirely in place. As prices fall, the market will grow exponentially, and vice versa.

The emergence of private access to space represents a monumental milestone in the trajectory of human civilization, with profound implications for societal transformation. This paradigm shift not only embodies a watershed moment in technological innovation but also carries the potential to engender a myriad of consequential impacts across various facets of human existence. The ramifications of this transformative trajectory extend far beyond the realm of mere commercial enterprise, permeating the fundamental fabric of societal dynamics and human consciousness. The burgeoning space tourism industry concurrently catalyzes the evolution of a diverse ecosystem of ancillary activities, engendering a symbiotic network of interconnected endeavors that collectively underpin and support this nascent frontier of human exploration.

Furthermore, as the once-fanciful prospect of space travel gradually materializes into a plausible reality, it precipitates a seismic and paradigmatic shift in human perception and consciousness. The profound experience of voyaging beyond Earth's confines engenders a transcendent phenomenon known as the Overview Effect, a cognitive transformation characterized by an intensified awareness of human interconnectedness and an acute understanding of the fragility and preciousness of our planet. The profound cognitive impact of the Overview Effect precipitates a potential reconfiguration of societal values and priorities, fostering a heightened collective consciousness pertaining to ecological stewardship and a redefined sense of global citizenship. This cognitive recalibration may foster a fundamental transformation in societal ethos, birthing a heightened imperative for global cooperation and environmental preservation, catalyzing a metamorphosis of societal paradigms and fostering an era of heightened socio-ecological consciousness and stewardship.

In the domain of space tourism, the integration of mixed frequency data heralds a pivotal juncture in the futuristic trajectory of travel demand prediction research. Although the utilization of mixed frequency data in predicting travel demand remains a relatively nascent endeavor, the integrative fusion of disparate data frequencies represents a promising avenue for future exploration and scholarly inquiry. The innovative incorporation of mixed frequency data holds the potential to engender an enriched and comprehensive predictive framework, capable of garnering nuanced insights into the complex dynamics of travel demand, thereby facilitating the formulation of more accurate and robust forecasting models.

As the era of big data continues to burgeon, the confluence of big data analytics and mixed frequency methodologies presents a promising frontier for research within the domain of travel demand forecasting. The integration of big data analytics contributes an expansive reservoir of multidimensional variables, furnishing a rich tapestry of empirical inputs that stand to fortify the predictive precision and anticipatory efficacy of travel demand models. The augmentation of travel demand forecasting through the assimilation of both big data and mixed frequency data methodologies holds immense promise in furnishing a more holistic and nuanced understanding of the intricate dynamics underpinning travel demand, thereby paving the way for a paradigmatic advancement in the scholarly praxis of space tourism research.

As the number of innovative models for travel demand forecasting increases, hybrid models incorporating new forecasting models will become the standard for the future to improve the accuracy of forecasts. Although spatial regression-based forecasting is rarely applied, spatial regression models can be considered an innovative approach to travel demand forecasting as they believe in geographical information. Wu et al. (2017) also use spatial regression models for hotel and tourism demand analysis. There are several applications in tourism economics. In the future, as the demand for space tourism rises, the above econometric models could also be used for predictive analysis.

References

Abdou, M., Musabanganji, E., & Musahara, H. (2021). Tourism demand modelling and forecasting: A review of literature. *African Journal of Hospitality, Tourism and Leisure, 10*(4), 1370–1393. https://doi.org/10.46222/ajhtl.19770720-168

Abrate, G., Nicolau, J. L., & Viglia, G. (2019). The impact of dynamic price variability on revenue maximization. *Tourism Management, 74*, 224–233. https://doi.org/10.1016/J.TOURMAN.2019.03.013

Albaladejo, I. P., González-Martínez, M. I., & Martínez-García, M. P. (2016). Nonconstant reputation effect in a dynamic tourism demand model for Spain. *Tourism Management, 53*, 132–139. https://doi.org/10.1016/J.TOURMAN.2015.09.018

Aviv, Y., & Pazgal, A. (2008). Optimal pricing of seasonal products in the presence of forward-looking consumers. *Manufacturing & Service Operations Management, 10*(3), 339–359. https://doi.org/10.1287/MSOM.1070.0183

Beery, J. (2012). State, capital and spaceships: A terrestrial geography of space tourism. *Geoforum, 43*(1), 25–34. https://doi.org/10.1016/J.GEOFORUM.2011.07.013

Bunghez, C. L. (2015). Space tourism market analysis: Current situation and future trends. *International Conference on Marketing and Business Development, I*(1), 97–103. http://www.mbd.ase.ro/?p=107

Cater, C. I. (2010). Steps to space; opportunities for astrotourism. *Tourism Management, 31*(6), 838–845. https://doi.org/10.1016/J.TOURMAN.2009.09.001

Chang, Y. W. (2015). The first decade of commercial space tourism. *Acta Astronautica, 108*, 79–91. https://doi.org/10.1016/J.ACTAASTRO.2014.12.004

Crouch, G. I. (1995). A meta-analysis of tourism demand. *Annals of Tourism Research, 22*(1), 103–118. https://doi.org/10.1016/0160-7383(94)00054-V

Duan, W. (2020). Ordinal logistic regression analysis on influencing factors of space tourism expectation model. *Journal of Physics: Conference Series, 1651*(1), Article 012066. https://doi.org/10.1088/1742-6596/1651/1/012066

Forganni, A. (2017). The potential of space tourism for space popularisation: An opportunity for the EU Space Policy? *Space Policy, 41*, 48–52. https://doi.org/10.1016/J.SPACEPOL.2017.04.005

Frost, J., & Frost, W. (2021). Exploring prosocial and environmental motivations of frontier tourists: Implications for sustainable space tourism. *Journal of Sustainable Tourism, 30*(9), 2254–2270. https://doi.org/10.1080/09669582.2021.1897131

Gabriel, J., Stefan, B., & Schubert, F. (2008). The economic effects of advertising on tourism demand. *Economics Bulletin, 6*(45), 1–16.

Goehlich, R. A. (2005). A Ticket pricing strategy for an oligopolistic space tourism market. *Space Policy, 21*(4), 293–306. https://doi.org/10.1016/J.SPACEPOL.2005.08.007

Goh, C., & Law, R. (2011). The methodological progress of tourism demand forecasting: A review of related literature. *Journal of Travel & Tourism Marketing, 28*(3), 296–317. https://doi.org/10.1080/10548408.2011.562856

Guizzardi, A., & Stacchini, A. (2015). Real-time forecasting regional tourism with business sentiment surveys. *Tourism Management, 47*, 213–223. https://doi.org/10.1016/J.TOURMAN.2014.09.022

Gunter, U., & Önder, I. (2015). Forecasting international city tourism demand for Paris: Accuracy of uni- and multivariate models employing monthly data. *Tourism Management, 46*, 123–135. https://doi.org/10.1016/J.TOURMAN.2014.06.017

Hsu, C. I., Li, H. C., & Yang, L. H. (2013). Intertemporal demand for international tourist air travel. *Transportmetrica a: Transport Science, 9*(5), 385–407. https://doi.org/10.1080/18128602.2011.581254

Jacobs, L., Du Preez, E. A., & Fairer-Wessels, F. (2019). To wish upon a star: Exploring Astro Tourism as vehicle for sustainable rural development. *Development Southern Africa, 37*(1), 87–104. https://doi.org/10.1080/0376835X.2019.1609908

Karl, M., Chien, P. M., & Ong, F. (2021). Impulse buying behaviour in tourism: A new perspective. *Annals of Tourism Research, 90*, Article 103136. https://doi.org/10.1016/J.ANNALS.2020.103136

Li, G., Song, H., & Witt, S. F. (2005). Recent developments in econometric modeling and forecasting. *Journal of Travel Research, 44*(1), 82–99. https://doi.org/10.1177/0047287505276594

Lin, V. S., & Song, H. (2014). A review of Delphi forecasting research in tourism. *Current Issues in Tourism, 18*(12), 1099–1131. https://doi.org/10.1080/13683500.2014.967187

Margolis, E. A. (2020). 'See your spaceport': Project Apollo and the origins of Kennedy Space Center tourism, 1963–67. *European Journal of American Culture, 39*(3), 249–274. https://doi.org/10.1386/EJAC_00030_1/CITE/REFWORKS

Moreau, C. P., Markman, A. B., & Lehmann, D. R. (2001). "What is it?" Categorization flexibility and consumers' responses to really new products. *Journal of Consumer Research, 27*(4), 489–498. https://doi.org/10.1086/319623

Morley, C. L. (2009). Dynamics in the specification of tourism demand models. *Tourism Economics, 15*(1), 23–39. https://doi.org/10.5367/000000009787536654

Olya, H., & Han, H. (2022). Emerging space tourism business: Uncovering customer avoidance responses and behaviours. *Journal of Vacation Marketing*. https://doi.org/10.1177/13567667221101408/ASSET/IMAGES/LARGE/10.1177_13567667221101408-FIG2.JPEG

Peng, B., Song, H., & Crouch, G. I. (2014). A meta-analysis of international tourism demand forecasting and implications for practice. *Tourism Management, 45*, 181–193. https://doi.org/10.1016/J.TOURMAN.2014.04.005

Reddy, M. V., Nica, M., & Wilkes, K. (2012). Space tourism: Research recommendations for the future of the industry and perspectives of potential participants. *Tourism Management, 33*(5), 1093–1102. https://doi.org/10.1016/J.TOURMAN.2011.11.026

Savino, R., Russo, G., Carandente, V., & D'oriano, V. (2013). Hyplane: Challenges for space tourism and business transportation. *Journal of Aerospace Engineering* 2(123), 2. https://doi.org/10.4172/2168-9792.1000123

Smeral, E., Witt, S. F., & Witt, C. A. (1992). Econometric forecasts: Tourism trends to 2000. *Annals of Tourism Research, 19*(3), 450–466. https://doi.org/10.1016/0160-7383(92)90130-H

Soleimani, S., Bruwer, J., Gross, M. J., & Lee, R. (2018). Astro-tourism conceptualisation as special-interest tourism (SIT) field: A phenomonological approach. *Current Issues in Tourism, 22*(18), 2299–2314. https://doi.org/10.1080/13683500.2018.1444021

Song, H., & Li, G. (2008). Tourism demand modelling and forecasting—A review of recent research. *Tourism Management, 29*(2), 203–220. https://doi.org/10.1016/J.TOURMAN.2007.07.016

Song, H., Li, G., Witt, S. F., & Athanasopoulos, G. (2011). Forecasting tourist arrivals using time-varying parameter structural time series models. *International Journal of Forecasting, 27*(3), 855–869. https://doi.org/10.1016/J.IJFORECAST.2010.06.001

Song, H., Qiu, R. T. R., & Park, J. (2019). A review of research on tourism demand forecasting: Launching the Annals of Tourism Research Curated Collection on tourism demand forecasting. *Annals of Tourism Research, 75*, 338–362. https://doi.org/10.1016/J.ANNALS.2018.12.001

Tasci, A. D. A., Fyall, A. D., & Fu, X. (2021). Social representations of space travel: Modeling the antecedents and outcomes. *International Journal of Tourism Research, 23*(4), 611–635. https://doi.org/10.1002/JTR.2430

Toivonen, A. (2020). Sustainability dimensions in space tourism: The case of Finland. *Journal of Sustainable Tourism, 30*(9), 2223–2239. https://doi.org/10.1080/09669582.2020.1783276

Tukamushaba, E. K., Lin, V. S., & Bwire, T. (2013). Modeling and forecasting inbound tourism demand for long-haul markets of Beijing. *Journal of China Tourism Research, 9*(4), 489–506. https://doi.org/10.1080/19388160.2013.841505

van Doorn, J. W. M. (1984). Tourism forecasting and the policymaker: Criteria of usefulness. *Tourism Management, 5*(1), 24–39. https://doi.org/10.1016/0261-5177(84)90005-0

Vanegas, M., Sr., & Vanegas, M. (2013). Co-integration and error correction estimation to forecast tourism in El Salvador. *Journal of Travel & Tourism Marketing, 30*(6), 523–537. https://doi.org/10.1080/10548408.2013.810992

Wan, S. K., & Song, H. (2018). Forecasting turning points in tourism growth. *Annals of Tourism Research, 72*, 156–167. https://doi.org/10.1016/J.ANNALS.2018.07.010

Witt, S. F., & Witt, C. A. (1995). Forecasting tourism demand: A review of empirical research. *International Journal of Forecasting, 11*(3), 447–475. https://doi.org/10.1016/0169-2070(95)00591-7

Wong, I. K. A., Fong, H. I. V., Leong, A. M. W., & Li, J. X. (2019). Predicting event tourists' gambling decision: The cross-level effect of brand equity. *International Journal of Contemporary Hospitality Management, 31*(7), 2951–2969. https://doi.org/10.1108/IJCHM-05-2018-0434/FULL/PDF

Wu, D. C., Song, H., & Shen, S. (2017). New developments in tourism and hotel demand modeling and forecasting. *International Journal of Contemporary Hospitality Management, 29*(1), 507–529. https://doi.org/10.1108/IJCHM-05-2015-0249/FULL/PDF

Chapter 5
What Are the Problems to Sutdy Space Tourism Demand Forecasting with Time Series Analysis?

1 Introduction

The tourism industry is not only an indispensable key sector in the gross domestic product of many regions, but also one of the industry's most sensitive to external crises and changes (Zhang et al., 2021). The industry is volatile, which directly affects the vitality and resilience of the local economy. Consequently, it becomes paramount for those engaged in tourism management to establish reliable forecasting models and provide precise estimates of tourist demand, enabling informed decision-making and strategic planning (Goh & and Law, 2011). These reliable forecasting data are not only fundamental to tourism planning and operation, but they also signal steady progress for the industry. At the operational level, accurate forecasts of tourist demand can assist companies in making rational decisions regarding resource allocation, streamlining service processes, ensuring the stability and immediacy of income sources, and thereby fostering the healthy development of the industry as a whole. For government officials, these reliable forecasting data are of particular importance when assessing global tourism market trends, formulating relevant policies, and developing response strategies (Álvarez-Díaz & Rosselló-Nadal, 2010). These data function as a compass for decision-making, aiding policymakers in capitalizing on emerging opportunities within a shifting global tourism landscape, formulating scientific plans, and fostering sustained prosperity for regional and even national economies.

Over the years, academics and industry professionals have pushed the boundaries of forecasting by introducing a variety of forecasting tools and techniques to overcome uncertainty in the tourism market (Song et al., 2023). One such tool is the use of the time series analysis to predict future tourism demand (e.g., arrivals, hotel bookings) based on historical patterns. Econometric models have also been employed to establish a correlation between tourism demand and economic and external variables (e.g., GDP, exchange rates, airfare). Machine learning has been employed to address complex, non-linear relationships and voluminous data sets, including social media trends and web traffic. For instance, random forests and XGBoost techniques have

K. Peng et al., *Space Tourism Supply Chain*, Contributions to Management Science,
https://doi.org/10.1007/978-981-95-1080-1_5

been used to forecast visitor numbers based on characteristics such as marketing spend, weather or events. Additionally, sentiment analysis has been adopted to infer travel intent from social media or reviews. Goh and and Law (2011) have noted that the accuracy of these forecasting methods is often assessed through a combination of evaluations, indicating that a single indicator is often unable to fully capture their effectiveness. For example, combining statistical and machine learning methods for improved accuracy. To further explore the state of the art in this field, Song et al. (2023) conducted an in-depth study in which they meticulously reviewed 211 influential papers on tourism market forecasting published between 1968 and 2018. This additional review reveals an interesting phenomenon: as the variety of forecasting models grows richer, the accuracy of forecasts shows an upward trend. This suggests that the tourism market's predictive capacity is steadily rising as technology progresses and models undergo further innovation. It is worth noting, however, that while the increasing variety of models has led to improved accuracy, there is no "universal" approach that is significantly better than other competing models in all cases. Each prediction model possesses a distinct scope and limitations, which must be evaluated and selected in specific application scenarios. Consequently, the advancement of tourism demand forecasting strategies continues to be an active and challenging research domain. It is incumbent upon scholars to perpetually explore novel forecasting instruments and methodologies, while concurrently refining extant models and methodologies to enhance the precision and dependability of forecasts. This systematic approach is instrumental in addressing the intricacies and unpredictability of the tourism market, thereby ensuring the provision of robust support for the sustainable advancement of the tourism industry.

In order to enhance the economic, fiscal, and social significance of tourism, there is a necessity for the development of research methodologies that address the collection and forecasting of tourism time data. Predicting the number of tourists in a specific time period is challenging due to the potential for unplanned changes to disrupt people's interest (Hao et al., 2020). Consequently, the accurate prediction of tourist arrivals at specific times is crucial. Accurate tourism demand forecasting is essential for policy making.

Space tourism, a highly cutting-edge concept in today's world, is gradually unveiling its mysterious veil. A close examination of the number of international tourists entering space reveals significant annual fluctuations, attributable to factors such as space safety, physical endurance, and numerous other unforeseen elements (Uğur & Akbıyık, 2020). Accurate prediction of the number of international tourists entering space is of vital importance to authorities and business participants. This facilitates the formulation of informed and strategic decisions, while concurrently reducing waste and inefficient use of tourist resources. Accurate prediction models empower stakeholders with the ability to plan and allocate their resources with greater precision, thereby reducing operational risks and uncertainties (Chen et al., 2015).

This study aims to pioneer a new trend in tourism demand forecasting, with a focus on advancing cutting-edge forecasting methods rooted in the highly regarded nonlinear "ARIMA Models" (Pankratz, 2008). The present study undertakes a dual-pronged approach that encompasses both an exhaustive theoretical exploration of these novel methods and a comparative analysis against established benchmark models that epitomize the state-of-the-art in parametric and non-parametric forecasting techniques. The effectiveness of these novel methods is then comprehensively evaluated. These achievements signify a significant milestone in the evolution of space tourism. As the industry continues to flourish and technology continues to innovate, it is reasonable to predict that the field of space tourism will continue to make new and significant contributions in the years and decades to come.

2 Literature Review

Space Tourism and Time Series Analysis

The concept of space tourism is predicated on the infinite human yearning for leisure, entertainment, and scientific exploration. Passengers, who pay a specified fee, board a spacecraft designed specifically for space travel, thereby embarking on an extraordinary journey through the vast universe. In the realm of space tourism, suborbital flight offers a remarkable prelude, providing passengers with an ephemeral yet profound experience of weightlessness that defies gravity. This unique vantage point offers a panoramic view of the Earth, affording passengers a breathtaking perspective that encompasses both the terrestrial landscape and the celestial expanse. Orbital flight constitutes a comprehensive exploration of space tourism. Passengers will have the opportunity to extend their stay in space and experience a more protracted journey through space. A particularly salient aspect of this experience is the opportunity for passengers to visit renowned space landmarks, such as the International Space Station (ISS), thereby gaining proximity to the vanguard of human space exploration. This unique advantage enables them to personally encounter both the marvels of technological achievement and the enigmatic nature of the universe. Space tourism is not merely a recreational activity; it is also a captivating "hobby." As the European Space Agency (ESA) (2008) has noted, space tourism relies on privately funded and operated aircraft to conduct suborbital flights and is committed to the continuous development of space propulsion technology. This commitment not only signifies an ongoing expansion of the boundaries of space exploration but also underscores the gradual realization of the popular aspiration to engage with and experience space.

In recent years, the advent of commercial aerospace enterprises and uninterrupted technological breakthroughs have led to a substantial reduction in the cost of space travel. Commercial aerospace companies, such as SpaceX, Blue Origin, and Virgin Galactic, have played a pivotal role in this transformation by adopting reusable rocket technology. Suborbital flight, as an entry-level project for space tourism, is gaining popularity. An examination of the contemporary state of space tourism market development reveals a period of accelerated growth. The continuous breakthroughs in commercial aerospace technology and the gradual reduction in costs have transformed space travel from being exclusive to the affluent to a feasible aspiration for a broader segment of the population. Concurrently, the enhancement of living standards and the diversification of tourism demands have led to an increased focus on space tourism as a novel and appealing travel option.

Time series analysis, formally referred to as TSA (Kočenda & Černý, 2015), is a statistical technique that utilizes raw data or historical statistics to generate facts and make evaluations. The technique involves a thorough examination of the statistical laws inherent in the data, with a particular emphasis on those that reveal cyclical fluctuations. The crux of this analysis lies in its ability to accurately predict and reasonably assess future trends within the designated field by identifying and examining these underlying laws. Specifically, time series analysis involves the study of chronologically ordered data sets that can represent various phenomena such as economic indicators, climate conditions, stock prices, etc. Analysts employ a range of statistical models and algorithms, including autoregressive models, moving averages, exponential smoothing techniques, and the more complex ARIMA models, to extract trends, seasonality, and cyclical components from the data.

From the perspective of application scenarios, time series analysis is widely used in many fields, including economic forecasting, financial market analysis, disease transmission monitoring, weather forecasting, and industrial quality control. In the field of economic, for instance, it aids in the prediction of pivotal economic indicators, including GDP growth rates and unemployment rates (Wen et al., 2022). Within the financial sector, it plays a pivotal role in stock price forecasting and risk management. Similarly, in the domain of public health, its significance lies in the prediction of epidemic development trends. The objective of this analysis is twofold: first, to unveil the dynamic patterns underlying the data, and second, to utilize these patterns to formulate forward-looking predictions and furnish a scientific foundation for decision makers.

The efficacy of time series analysis stems from its ability to process data that undergoes changes over time. It effectively captures and explains the time dependencies in the data, thereby producing more accurate and timely forecast results in comparison to other non-time series methods. Furthermore, by identifying periodic adjustments, it facilitates comprehension and prediction of the impact of periodic events on the system, thereby enhancing the ability to discern future trends (Dong et al., 2023).

The assessment of the smoothness of time series constitutes a pivotal element within the broader framework of time series analysis. The objective of this test is to ascertain whether the data possesses stationary characteristics. That is to say, it is necessary to determine whether the statistical properties of the data (e.g., mean, variance, etc.) remain constant over time. The outcomes of this test are pivotal in selecting an appropriate modeling technique. In instances where the data is found to be non-stationary, it is often imperative to stabilize it through the implementation of difference or other transformation methods, thereby ensuring the efficacy of subsequent modeling and prediction endeavors (Tsay & Chen, 2018).

According to the handbook of time series analysis (Cang & Seetaram, 2012), the ARIMA (p, d, q) model is a widely employed time series data modeling technique that aims to elucidate and predict patterns in time series data. The acronym

"ARIMA" stands for "Autoregressive Integrated Moving Average." In this model, p, d, and q represent the order of the autoregressive term, the number of differences, and the order of the moving average term, respectively. The model integrates the strengths of autoregression (AR) and moving average (MA) by processing non-stationary data through differences (I, i.e., d-time differences). The ARIMA model has the capacity to identify trends, seasonality, and cyclical components in time series, thereby producing more accurate forecasting results. Specifically, when d = 0, the ARIMA (p, 0, q) model is simplified to the ARMA (p, q) model. The ARMA model finds primary application in the explanation and prediction of patterns in stationary data, such as financial time series. The model's architecture encompasses two fundamental components: an autoregressive term, which encapsulates historical information within the data, and a moving average term, which serves to mitigate the impact of random error terms, thereby enhancing the predictability of the model's outputs. The four primary components of time series analysis generally comprise:

Data Preprocessing: This involves a variety of activities, including data cleaning, handling missing values, outlier detection and processing, and other related tasks. The primary objective of these activities is to ensure the accuracy and reliability of the data.

Model Selection: select a model that is commensurate with the characteristics of the data and the objective of the analysis, such as an ARIMA model or an exponential smoothing model.

Estimation of Model Parameters: achieved through the employment of statistical methodologies to calculate the parameters within the model, such as autoregressive coefficients and moving average coefficients.

Forecasting and Evaluation: employ the estimated model to make predictions and evaluate the accuracy and reliability of the prediction results.

In the study by Bi et al. (2021), the researchers first conducted a smoothness test on the time series to ascertain the stationarity of the data. Subsequently, they implemented the ARIMA (p, d) model for modeling and employed the estimated results of the model for forecasting. This approach integrates a series of critical steps and components of time series analysis, ensuring comprehensive and precise analysis.

In the domain of time series analysis, the least squares method is a technique employed for estimating model parameters. This method is frequently utilized in the fitting of linear models, such as the ARIMA model, an acronym for the autoregressive integrated moving average model, involves the differentiation of the "integral" component to achieve stationarity. When implementing an ARIMA model, the least squares method is employed to estimate the model parameters (e.g., the autoregressive and moving average coefficients) with the objective of minimizing forecast errors.

Time series analysis and regression analysis are two important methods for building mathematical models to represent and predict statistical patterns. In the context of TSA, prominent tools include ARIMA models, LSTM networks, and regression prediction models such as XGBoost (extreme gradient boosting) and SVM (support vector machines). The selection of these models and methods is contingent upon the characteristics of the data and the particular requirements of the prediction task at hand (Carmona-Benítez & Nieto, 2020).

In TSA, mathematical models are constructed to encapsulate the time-dependent characteristics of data. These models are then employed to forecast future values. A prominent example of a linear model employed in TSA is the ARIMA model, which encapsulates trend, seasonality, and randomness in time series through a combination of autoregressive, differencing, and moving average components. In addition to the ARIMA model, the Long Short-Term Memory (LSTM) network is another model that is widely used in TSA, especially when dealing with time series data with long-term dependencies. LSTM represents a distinctive form of recurrent neural network (RNN) that addresses the gradient vanishing or exploding problem traditionally encountered by conventional RNNs when processing extensive sequence data.

Regression analysis is a statistical method that is employed to study the relationship between one or more independent variables (predictors) and a dependent variable (response variable). The objective of regression analysis is to construct a mathematical model that can articulate the statistical relationship between independent and dependent variables. This model is then employed to predict the value of the dependent variable. In the field of regression analysis, a multitude of models can be employed, including linear regression, polynomial regression, and logistic regression.

Furthermore, machine learning algorithms, including XGBoost and SVM, can also be employed for regression analysis tasks, particularly when the data relationship is nonlinear or complex. In TSA, regression forecasting models such as linear regression, XGBoost, SVM, and LSTM can be utilized to predict future values of time series. These models have the capacity to identify trends and patterns within the time series, leveraging these patterns to inform predictions. However, it is imperative to acknowledge that the efficacy of these models is contingent upon the nature of the data and the specific problem being addressed. To illustrate this point, consider the following examples: LSTM may be more suitable for processing time series data with long-term dependencies, while XGBoost may be more suitable for processing data with nonlinear relationships.

The enterprise of space tourism is still in its infancy. The dearth of historical data hinders the execution of time series analysis. Consequently, the present study endeavors to furnish alternative approaches to address the dearth of data in space tourism research, with a particular emphasis on three aspects of extant literature: smoothness testing, ARIMA (p, d) modeling, and forecasting.

Figure 1 is the FAA Commercial Space Data from 2010 to 2023. The graphic summarizes the Federal Aviation Administration's annual report on commercial Space Transportation, including information on licensed launch sites, airspace, launch and reentry activities, major operators and launch sites, and temporal and geographic distribution. Despite the relatively weak direct correlation between FAA commercial space data and tourism demand, in practical applications, a comprehensive consideration of multiple factors and data sources enables exploration of its potential impact and relationship from a broader perspective, facilitating accurate

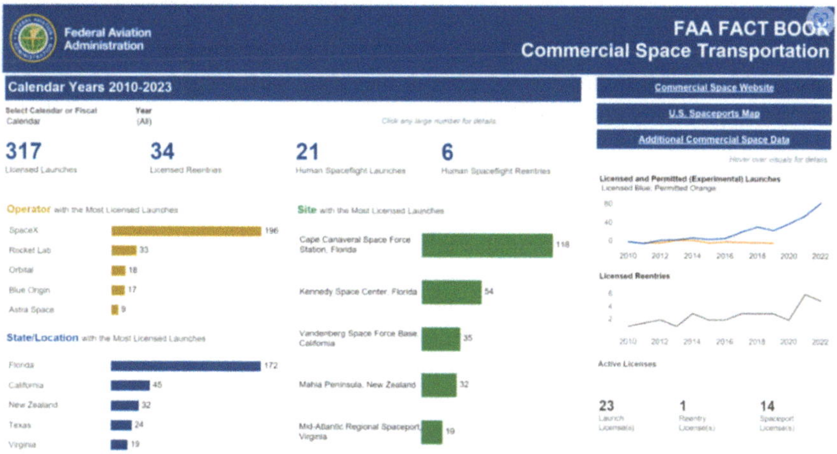

Fig. 1 Federal aviation administration statistics of commercial space transportation

predictions and decisions. Given the inherent temporal progression exhibited by this time series, it is more persuasive to select this time series as the object of this study.

Space Tourism Market Forecasting and Planning

Space tourism is transitioning from the realm of science fiction to that of reality, driven by substantial technological advancements and policy support, which have transformed it into a rapidly expanding field. A series of disruptive technological innovations have enhanced the safety and feasibility of space tourism, establishing a substantial foundation for interstellar travel. In response, governments worldwide have enacted policies and measures to foster the growth of this emerging industry. These policies encompass a wide range of domains, including technology research and development, infrastructure construction, security measures, and market access. The comprehensive nature of these policies provides substantial guarantees for the healthy development of the space tourism industry.

Presently, the development of the overall space economy (comprising satellite communications, deep space exploration, etc.) possesses the potential to generate revenue amounting to one trillion dollars. Although the independent scale forecast of space tourism as a subset is still unclear, the accelerating commercialization of suborbital travel and technological progress in orbital travel indicate that the market has huge growth potential. For instance, Virgin Galactic has received hundreds of tourists, with ticket prices reduced from millions of dollars to 200,000–450,000 US dollars, with a flight time of about 1 h and a few minutes of weightlessness. Chinese companies, such as Deep Blue Aerospace, have announced plans to launch reusable aircraft by 2028, with ticket prices projected to range from 2 to 3 million RMB.

SpaceX's Starship, a reusable spacecraft, has been developed to reduce the cost of orbital travel by more than 50%. This vehicle has also conducted all-private manned space missions. The future development of the space tourism destination market is expected to primarily encompass ground simulation experiences, suborbital flights, and orbital flights.

The tourism industry is experiencing a paradigm shift, driven by the emergence of space tourism. It disrupts the geographical boundaries of conventional tourism, redefining the concept of "remoteness." This transformation extends the scope of travel experiences, allowing people to travel from the surface of the earth to the depths of the universe, and provides tourists with an unprecedented visual shock and emotional touch. This transformative travel experience not only enhances the diversity of tourism products but also fosters global enthusiasm and yearning for space exploration, thereby generating a steady stream of passenger flow and market attention. Moreover, the emergence of space tourism has catalysed the advancement of related technologies, spurred the extension and modernisation of associated industrial value chains, and generated employment opportunities and economic benefits within the tourism sector. Concurrently, space tourism has also promoted the innovation and optimization of tourism services, enhancing the overall competitiveness and brand influence of the tourism industry.

According to the most recent Space Market Forecast report (Grand View Research, 2024; openPR, March 2025), the market is undergoing substantial growth due to technological advancements, an increase in interest in adventure travel, and high-net-worth individuals who are interested in orbital flights. Technological innovations, including reusable rockets and advanced spacecraft designs, have rendered astronomical travel increasingly feasible and cost-effective. This, in turn, has been further fueled by the growing demand from affluent individuals seeking unique experiences. Consequently, the industry is undergoing a period of substantial expansion.

Grand View Research forecasted the trends in the space tourism market from 2024 to 2030, based on data from 2017 to 2023, titled "Space Tourism Market Size, Share and Trends" (July 2024). The study has segmented the global space tourism market by type, end-use, and region. In terms of type, the market was segmented into orbital, suborbital, and others. The suborbital segment dominated the overall market in 2023 with a market share of 48.5% and was expected to grow at a Compound Annual Growth Rate (CAGR) of more than 44.4%, while the orbital segment was expected to achieve the fastest CAGR of 45.5%. Based on end use, a total of 13 commercial spaceflight missions were conducted by private and government organizations in 2021, of which 7 were successful. The market was segmented into government, commercial, and other segments. The commercial segment was expected to dominate with a market share of 55.8% by 2023. The market was expected to grow at a CAGR of 45.3% during the forecast period, which was the fastest growth rate among all the segments. In addition, the government segment was expected to grow at a healthy CAGR of 44.6% during the forecast period. From a regional perspective, the North American space tourism market dominated the global landscape in 2023, commanding the largest revenue share of 38.9%. The region boasts a highly developed infrastructure and a robust R&D base, positioning it as the leading contributor to the global market during the forecast period. Within the United States, the space tourism market accounted for the largest revenue share of 83.1% in the North American region in 2023 and was expected to maintain its dominance throughout the forecast period. This trend was attributable to two key factors: the preference for space tourism among high-net-worth individuals and the region's increasing technology investments. Furthermore, advancements in space travel technologies, such as reusable rockets and spacecraft, which have led to a reduction in the cost and enhancement in the accessibility of space tourism experiences, were anticipated to further stimulate the growth of the U.S. market. Advancements in aerospace technology, augmented private sector participation, and government support were also driving factors. The European space tourism market was expected to demonstrate a substantial CAGR during the forecast period from 2024 to 2030. In 2023, the Russian space tourism market accounted for the largest revenue share in the European region, primarily due to its rich space heritage, relatively affordable space tourism options, and extensive space infrastructure and launch capabilities. The space tourism market in Asia Pacific was also expected to undergo significant growth during the forecast period, particularly at a CAGR of 45.1% between 2024 and 2030. The governments of China, India, and Japan have been allocating substantial resources to the development of space exploration and tourism programs. The CNSA, the Indian Space Research Organization (ISRO), and the Japan Aerospace Exploration Agency (JAXA) are the primary contributors to the growth of the Asia Pacific market. A growing interest in space exploration and tourism has been observed across the Asia Pacific region, driven by economic growth and rising disposable income, which has led to an increased demand for novel and adventurous experiences, including space travel. This expanding demand for space tourism presents significant opportunities for businesses and entrepreneurs to enter and innovate within the market.

According to Grand View Research, the number of interplanetary launches in 2023 was 223, including 109 from the United States, 67 from China, and 19 from Russia. It is anticipated that this figure will continue to increase in subsequent years. The global space tourism market size in 2023 was estimated at $888.3 million, with a projected valuation of $1.094 billion and a revenue of $10.09 billion in 2024. Projections indicate that the market will undergo a CAGR of 44.8% from 2024 to 2030. Another research company, OpenPR (March 2025), released a space tourism market forecast report that estimated the space tourism market size to be $1.2 billion in 2024. Projections indicate that, from 2026 to 2033, the market will expand at a CAGR of 25.4%, reaching a valuation of $8.5 billion by the year 2033.

The space tourism industry is characterized by a high degree of dynamism, attributable in large part to the considerable technological innovation it has experienced, as well as to the numerous mergers and acquisitions that have occurred within it. Propelled by technological advancements and substantial private investments, the space tourism market is undergoing a transformation, evolving from a niche luxury industry to a more expansive commercial sector. Presently, suborbital flights represent the predominant segment of the space tourism sector. However, it is anticipated that orbital and lunar tourism will emerge as pivotal drivers of future expansion.

Notwithstanding the challenges posed by cost and regulatory factors, advancements in reusable rocket technology and the emergence of international cooperation initiatives, such as China's Tiangong Station, point to the prediction of a market valuation exceeding US$30 billion by the year 2040 (openPR, March 2025).

According to CNSA white papers and CMSA announcements (www.cnsa.gov.cn), China successfully completed the Tianhe core module in 2021 and added two experimental modules (Wentian and Mengtian) in 2022, thereby finalizing the T-shaped basic configuration. The space station is currently fully operational, accommodating rotating crews of three astronauts for missions lasting six months. As of 2025, the station has conducted over 180 scientific experiments in the domains of microgravity physics, space life sciences, and technology demonstrations. The development of a next-generation crewed spacecraft that is reusable is underway for lunar missions. By the year 2030, China aspires to achieve a crewed lunar landing using the Long March-10 rocket and the Mengzhou spacecraft. A fourth group of astronauts, including payload specialists from Hong Kong and Macao, has been selected. The space station facilitates global research, with plans to host additional international experiments. Post-2025: The focus will shift to lunar missions and expanding space station utilization, including commercial cargo services (e.g., Qingzhou spacecraft).

China has achieved noteworthy advancements in space exploration, engaging in a multitude of experiments and conducting meticulous technical assessments. The country's space station is expected to remain in orbit for an extended period. Concurrently, space tourism has materialized as a tangible reality for a select demographic of affluent individuals. A market study (Wan, 2021) reveals that compared to young consumers in the United States, the younger generation in China has demonstrated a stronger interest in space tourism and space accommodation. The data revealed that Chinese respondents are willing to pay generously for a round-trip space trip to Earth, paying an average of about $1.6 million. In contrast, American respondents demonstrated a more conservative spending propensity, with an average expenditure of approximately $900,000. The space tourism industry is experiencing a period of significant growth, driven by the aspirations of a new generation of individuals who harbor dreams of space travel. This industry is poised to undergo rapid expansion on a global scale, offering a wealth of development opportunities that have never before been seen. Concurrently, the expanding global population and the sustained economic development are propelling individuals to augment their investment in space travel research. These investments, ranging from the exploration of cosmic mysteries to the pursuit of the unknown, are poised to significantly boost the demand for space tourism, propelling us into a new era of extensive space exploration.

At the national strategic level, China has formally adopted the space economy as a national strategy. The objective is to attain a space economy valued at $1.8 trillion by the year 2035, exhibiting an average annual growth rate of 9%. As a fundamental element of the "distance economy," the sector of space tourism is poised to benefit from policy support, technological innovation, and market demand growth. Furthermore, China intends to host the International Astronautical Congress (IAC) in 2035, a significant platform for showcasing the nation's space capabilities. By showcasing the Chang'e-6 lunar samples and manned lunar landing programs, China aims to attract international capital and technical cooperation, thereby promoting the global layout of space tourism. The Innovation of Business Models is a topic of significant concern to the IAC. A taxonomy of commercial space models reveals two distinct categories. The first category comprises models that are driven by the state-led space launch market. The second category comprises models that are dominated by private space companies. These models have gained significant traction within the space industry. However, it should be noted that China has not yet developed a mature commercial space model. To address this gap, it is recommended that the focus be directed towards the development of a space tourism business model within the context of the state-led space launch market. Given the developmental needs, coupled with the prevailing resource conditions and economic and technological capabilities, it is imperative to prioritize the promotion of manned spaceflight and lunar exploration initiatives as the cornerstone of commercial space development.

The initial phase in the development of a space tourism enterprise entails the formulation of forecasts. This process involves the analysis of market demand, the prediction of future trends in space tourism development, the design of space layouts, and the planning of directions and timetables for the subsequent decade. The ensuing decade is poised to be a period of accelerated growth for space tourism, both domestically and internationally. However, this period will also be accompanied by the accumulation of risks and problems as the space tourism industry undergoes a rapid development phase. A prevailing concern within the industry is the safety of space tourism. This concern is primarily attributable to the limited understanding of the universe that humans currently possess. Moreover, environmental concerns, exacerbated by global climate change and natural disasters, have profoundly altered the ecological dynamics of space. As human beings move into a new business and social paradigm driven by digitalization, automation, and other technologies, they may have elevated expectations for the technology of space tourism. Consequently, a significant challenge confronting the space tourism sector in the ensuing decade pertains to the management of technical and safety concerns. According to Crouch (2001), who examined the space tourism market using limited data, there were early warning signs. Goehlich (2005) explored the price tag pricing method for the space tourism market, which was first proposed in the subsequent three case studies, and its results have been listed. Furthermore, qualitative and quantitative inquiries provide a suitable foundation for comprehensive space tourism research.

3 Methodology

Database

In its most fundamental form, it is a style of aerial transportation. This study investigated that the general public of clients are folks who fly often, and then we can conclude that these customers tour about once every five years on average. So, the following action we can take is to classes our users. This study determined whether or not they had been to space in the closing five years based on the date of their most recent voyage. On the other hand, most humans on Earth have no longer been to space. In this scenario, this study used techniques from time series analysis to categorize each patron to decide whether they were allowed to go to the space area. We want to categorize the customers to show that the person organization in this scenario is not always foreseeable regarding prospective calls for using approaches primarily based on time series evaluation.

To make destination market projections, this study utilized the Arima version. To achieve this, the results categorized the customers and then computed the predicted destination market call for space tours that each institution of customers can have.

The capability to make predictions is a crucial part of time collection evaluation. When the prediction findings are analyzed, it is feasible to establish the general route in which things will broaden inside the destination market. This enables appropriate preventative movements to be made, which is essential for reaching the goal of preventing damage.

We looked for the following vital databases: Science Direct, Web of Science, Scopus, EBSCOhost, ProQuest Central, and China National Knowledge Network CNKI, as we can see that most authors' papers have close relationships with other space tourism authors. Figure 2 illustrates the development of space tourism since 2008. We can find that there are more space tourism research after 2020 with more connections between scholars. After searching the website "Web of Science" for the keywords: "space tourism," keywords are also mover relevant to space tourism research (see Fig. 3) in the last five years.

Using Time Series Analysis for Space Travel Demand Forecasting

Accurate forecasting becomes increasingly essential as the demand for the space area tour grows. As space tour becomes more on hand to the general public, traditional methods of tourism demand forecasting will not be enough to predict and meet the developing call for such specific reports.

One capacity approach for forecasting area travel demand might be via time series analysis. TSA is a statistical approach that models time-established statistics to identify patterns and tendencies. Search engine records are a usually used related variable for time-collection forecasts. Applying time collection analysis to space journey demand should involve reading web search phrases and online booking trends for area travel. Furthermore, researchers may want to check out social media information related to area travel or posts about area tourism experiences. This technique no longer most straightforwardly offers well-timed and frequent statistics but also captures potential space visitors' behaviors.

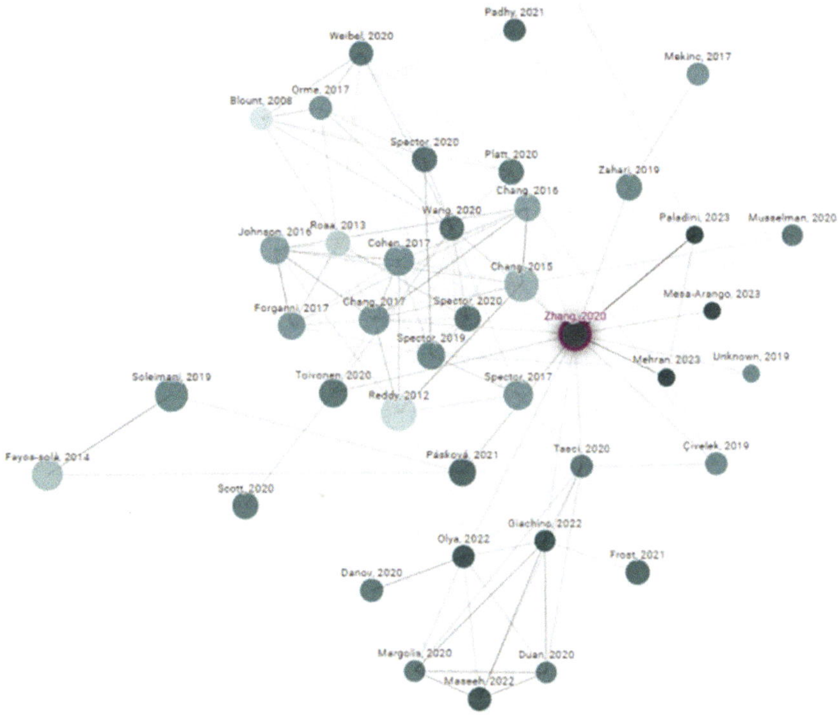

Fig. 2 Space tourism database connections

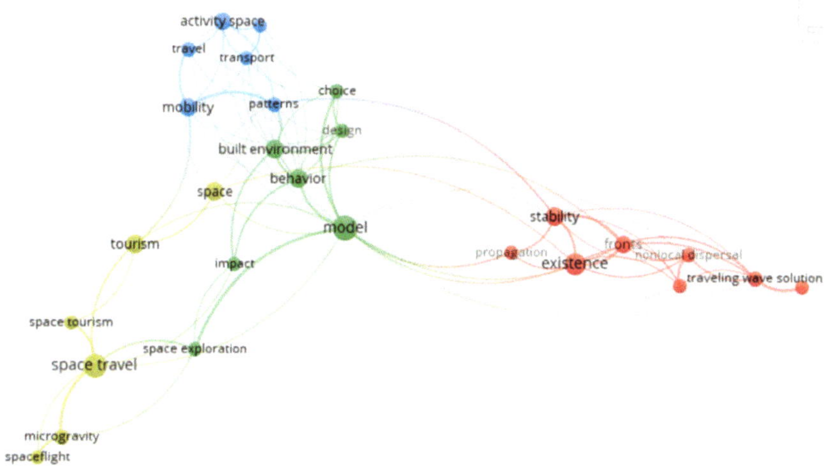

Fig. 3 Space tourism studies

This approach may want to provide perception into the purchaser's hobby and call for space-area travel, enabling tourism corporations to make facts-driven decisions about their advertising techniques and offerings (Dergiades et al., 2018). Moreover, this method may want to offer extra correct and up-to-date records compared to the traditional tourism call for forecasting strategies that depend on published statistical information with the behind-schedule guide (Lim et al., 2020) investigated the effect on arrivals to Korea from China, Hong Kong, Ind. The fundamental objective was to increase reliable and accurate predictive models for general global arrivals in Sri Lanka and its top 10 international procurement sites, the use of the SARIMA (Seasonal Autoregressive Integrated Moving Media) technique and month-to-month arrival facts (Senadeerage, 2020). Along with India and Taiwan, Japan, Thailand, and South Korea are China's top tourist destinations, assessing the results of large records and monetary factors. We searched the website Web of Science (WoS) for the keywords: "Time Series Analysis" and "Space Travel Demand Forecasting".

4 Results and Propositions

Results showed that time series analysis in space tourism is rare. The first reason is that the space tourism industry is nascent (Crouch et al., 2009; Peng et al., 2023), and the very limited data have been collected for model testing using time series analysis. It awaits more time for the space market to generate enough empirical data to test hypotheses and models of space tourism (Crouch, 2001). However, we can still summarize some finding in 4.1. The second reason is that the time series analysis is improperly applied to study space tourism behaviors. Many abnormal space tourism economic behaviors remain unexplained without essential space attributes like the measurement scale: spacetime. We elaborate two propositions in 4.2.

Findings in Literature

Time series analysis is the process of examining and forecasting data that undergoes changes over time. It is a statistical technique that involves the assessment of data quality. This assessment typically involves the examination of the data for trends, seasonality, and random components. However, the application of this technique is not limited to situations where there is sufficient time series data. In circumstances where data is limited, the concept of smoothness testing can be broadly applied to data quality assessment. For instance, one can ascertain whether the available data manifests discernible patterns or trends, even if these patterns do not strictly conform to time series trends. This assessment can be facilitated through the utilization of rudimentary visualization techniques, such as the employment of line charts, or the implementation of statistical assessments, including the application of moving averages or exponential smoothing to non-time series data. These methods are designed to identify underlying regularities in the data, although they may not attain the same degree of precision as time series analysis.

The ARIMA model is a widely utilized tool in the field of time series analysis, particularly for the purpose of capturing autoregressive, differencing, and moving average components in data. However, in instances where the availability of time series data is limited, the direct application of the ARIMA model may not be feasible. Consequently, alternative or variant formulations of the ARIMA model can be considered (Tsay & Chen, 2018):

Non-time series regression models: These models employ linear regression, polynomial regression, or machine learning regression models (e.g., random forests, gradient boosting machines) to construct forecasting models. These models have the capacity to predict the development of the space tourism industry based on non-time series characteristics (such as market size, technological progress, policy environment, etc.).

Simulation and scenario analysis: In instances where data is inadequate, simulation models can be utilized to generate data and explore various scenarios based on assumptions and expert judgment. Scenario analysis can facilitate decision makers' comprehension of potential outcomes under diverse conditions.

Bayesian methods: Bayesian statistics provides a methodology for integrating prior knowledge and new data, enabling inferences to be made even in scenarios where data is limited. Specifying prior distributions for model parameters and updating these distributions using Bayes' theorem results in more robust forecasts.

In the domain of space tourism, the practice of forecasting holds particular significance, as it enables stakeholders, including businesses, investors, and policymakers, to make informed decisions. While time series analysis is not a viable option, alternative methods exist for forecasting. For instance, the knowledge and experience of industry experts and market research data can be leveraged to predict the development trend of the space tourism industry. One such approach involves identifying analogous industries with ample historical data (e.g., the aviation and tourism industries) and leveraging their experience and trends to forecast the future of the space tourism industry. Furthermore, the integration of forecasts derived from multiple methodologies can enhance the accuracy and reliability of predictions. One such approach involves the utilization of machine learning algorithms, which have been demonstrated to effectively amalgamate the forecasts derived from disparate models.

Spacetime Propositions

Based on the literature and General Relativity Theory, we could develop propositions that refer to space attributes and observable phenomena of space economic behaviors measured by spacetime.

Proposition 1: Space tourism economic behaviors are determined by the nonzero curvature of spacetime.

Proposition 2: Earth tourism economic behaviors are determined by zero curvature of spacetime.

The two propositions argue that the difference of Space and Earth tourism economic behaviors are determined by the nonzero and zero curvatures of spacetime. However, Earth tourism is a special condition of zero curvature of spacetime, called flat space (Salman et al., 2022), that we have constructed current theories. The uniqueness of space tourism economic behaviors exists due to the sufficient condition of nonzero spacetime curvature. The implications are that a breakthrough cognition is needed to know and apply a sapcetime scale in space tourism studies. From the supply perspective, Earth tourism uses land resources as destinations in the production function to serve and predict inbound and outbound or rural and urban tourists (Kytzia et al., 2011). However, according to the outer space law, space does not belong to any territories/lands (Peterson, 1997), which means the production factors will be redefined in space economics as described in Schumpeter's innovation theory (Williams et al., 2021). From the demand perspective, Crouch (2001) indicated that motivation theory is probably one of the crucial questions in space tourism studies. The motivation for adventure tourism evident in people's demand depends on the prices offered and product attributes. The market equilibrium depends on supply and demand, which reshapes the market structure in terms of its dynamics and competition in space tourism (Crouch et al., 2009). The invariant of economic science is still the equilibrium of supply and demand, which can be extended from zero to nonzero curvature of spacetime. Measures the utility or trade-off between various product attributes, such as trip duration, comfort, activities, safety, and other features, is still the core of demand (Crouch, 2001; Crouch et al., 2009). The production function of space attributes would still follow the cost-benefit principles. Both together attain the price equilibrium of space travel. It is expected that there will be innovative theories for space tourism beyond the Karman line. Earth tourism within the Karman line continues the incremental theories. Various studies examine the market from alternative directions and methodologies so that the weight of evidence arising from these studies is consistent, mutually supportive, and therefore sufficiently compelling (Crouch, 2001).

5 Conclusions

Research in this field remains in its infancy, mainly in concept definition, enterprise improvement and financial outcomes, participating parties and affects, and environmental impacts; numerous controversies must be explored. However, area tourism is a desirable space industry.

Modern research in space tourism specializes in the sector as a whole instead of the specifics of its increase or the influences it may have from a micro viewpoint. This approach that most people of the studies on the perception of space tourism stays at the macro stage. Academics have lately begun paying greater attention to the issue of space tourism because of the implementation of pertinent country-wide rules, legal guidelines, and regulations; nonetheless, there is still a massive amount of debate over the suitable meanings of the term. There has been little or no examination of space tourism commercialization. In the destination market, more emphasis must be placed on studies into area tourism, which need to coincide with adopting and refining appropriate rules and methods.

A crucial element of commercial area tours is the burgeoning enterprise of space tourism, an emergent financial integral to expanding the distance economy. On the other hand, by using the superior era of spaceflight to the development of space tourism and selling the industrialization of space tourism, the age may be transformed and applied to attain monetary advantages with technological innovation. On the other hand, by using the generation to improve space area tourism and promote the industrialization of area tourism, the technology may be implemented to achieve business profits. As a result of the one-of-a-kind nature of the distance tourism industry, it will no longer be in direct opposition with other kinds of tourism or merchandise available on Earth. As a result, the destination market development of the space tourism enterprise seems promising and is expected to bring considerable economic advantages. This newsletter/book/research? (Oh now I know maybe newsletter is OK) aims to aid area applications and forecast future space tour calls based on a time collection evaluation.

There is now more debate than consensus over the primary players involved in space tourism regarding the topics who will be engaged (government, commercial organizations, schools, and social organizations). As the space tourism sector continues to expand and mature, the primary players will become more proactive in participating in space area tourism.

The dream of space area travel is no longer a fantasy beyond the reach of the general public. Various indications show that space travel could become popular within 15–20 years and become an affordable leisure activity. Technology and space exploration advancements have allowed individuals to experience suborbital flights, moon and Mars tours, or any other space exploration adventures. However, with the growing accessibility and affordability of space travel, there is a need for more accurate forecasting methods to meet demand.

In phrases of the impact on the environment, there is now minimal look at the environmental effects of area tourism, and there must be an increase in the number of studies on the impact on the surroundings in the destination market. At the same time, it is vital to understand that the emergence of area tourism as a brand-new business has had a constructive effect on the growth of the distance industry in all nations; consequently, it is vital that international collaboration and information sharing in this vicinity be improved.

References

Álvarez-Díaz, M., & Rosselló-Nadal, J. (2010). Forecasting British tourist arrivals in the Balearic Islands using meteorological variables. *Tourism Economics, 16*(1), 153–168. https://doi.org/10.5367/000000010790872079

Bi, J.-W., Li, H., & Fan, Z.-P. (2021). Tourism demand forecasting with time series imaging: A deep learning model. *Annals of Tourism Research, 90*, 103255. https://doi.org/10.1016/j.annals.2021.103255

Bradley, E., & Kantz, H. (2015). Nonlinear time-series analysis revisited. *Chaos: An Interdisciplinary Journal of Nonlinear Science, 25*(9). https://doi.org/10.48550/arXiv.1503.07493

Cang, S., & Seetaram, N. (2012). *Time series analysis: In handbook of research methods in tourism.* Edward Elgar Publishing.

Carmona-Benítez, R. B., & Nieto, M. R. (2020). SARIMA damp trend grey forecasting model for airline industry. *Journal of Air Transport Management, 82*, 101736. https://doi.org/10.1016/j.jairtraman.2019.101736

Chen, R., Liang, C.-Y., Hong, W.-C., & Gu, D.-X. (2015). Forecasting holiday daily tourist flow based on seasonal support vector regression with adaptive genetic algorithm. *Applied Soft Computing, 26*, 435–443. https://doi.org/10.1016/j.asoc.2014.10.022

Crouch, G. I. (2001). The market for space tourism: Early indications. *Journal of Travel Research, 40*(2), 213–219. https://doi.org/10.1177/004728750104000212

Crouch, G. I., Devinney, T. M., Louviere, J. J., & Islam, T. (2009). Modelling consumer choice behaviour in space tourism. *Tourism Management, 30*(3), 441–454. https://doi.org/10.1016/j.tourman.2008.07.003

Dergiades, T., Mavragani, E., & Pan, B. (2018). Google Trends and tourists' arrivals: Emerging biases and proposed corrections. *Tourism Management, 66*, 108–120. https://doi.org/10.1016/j.tourman.2017.10.014

Dong, Y., Xiao, L., Wang, J., & Wang, J. (2023). A time series attention mechanism based model for tourism demand forecasting. *Information Sciences, 628*, 269–290. https://doi.org/10.1016/j.ins.2023.01.095

Edgell, D. L. (2021). The worldly travelers of yesteryear contributed substantially to important tourism policy issues of today. *Journal of Hospitality & Tourism19*(1), 66–77.

European Space Agency. (2008). *Space tourism.* Retrieved January 15, 2022 fromwww.esa.int/esapub/bulletin/bulletin135/bul135c_galvez

Goehlich. (2005). A Ticket pricing strategy for an oligopolistic space tourism market. *Space Policy, 21*(4), 293–306. https://doi.org/10.1016/j.spacepol.2005.08.007

Goh, C., & Law, R. (2011). The methodological progress of tourism demand forecasting: A review of related literature. *Journal of Travel & Tourism Marketing, 28*(3), 296–317. https://doi.org/10.1080/10548408.2011.562856

Grand view research. (July 2024). *Space tourism market size, share & trends analysis report.* Retrieved March 29, 2025, from https://www.grandviewresearch.com/industry-analysis/space-tourism-market-report

Hao, F., Xiao, Q., & Chon, K. (2020). COVID-19 and China's hotel industry: Impacts, a disaster management framework, and post-pandemic agenda. *International Journal of Hospitality Management, 90*, Article 102636. https://doi.org/10.1016/j.ijhm.2020.102636

Kočenda, E., & Černý, A. (2015). *Elements of time series econometrics: An applied approach.* Charles University in Prague.

Kytzia, S., Walz, A., & Wegmann, M. (2011). How can tourism use the land more efficiently? Model-based approach to land-use efficiency of tourist destinations. *Tourism Management, 32*(3), 629–640. https://doi.org/10.1016/j.tourman.2010.05.014

Lim, S., & Giouvris, E. (2020). Tourist arrivals in Korea: Hallyu as a pull factor. *Current Issues in Tourism 23*(1), 99–130. https://doi.org/10.1080/13683500.2017.1372391

Mesa-Arango, P.-J., & J., Araujo, D. S. A., Bi, J., Basva, M., & Viti, F. (2023). Missions and factors determining the demand for affordable mass space tourism in the United States: A machine learning approach. *Acta Astronautica, 204*, 307–320. https://doi.org/10.1016/j.actaastro.2023.01.006

OpenPR. (2025, March). *Space tourism market exploring current trends and growth status for 2025.* Retrieved March 29, 2025 from https://www.openpr.com/news/3930831/space-tourism-market-exploring-current-trends-and-growth

Pankratz, A. (1983). *Forecasting with univariate Box-Jenkins models: Concepts and cases.* Wiley.

Pankratz, A. (2008). *Forecasting with univariate Box-Jenkins models: Concepts and cases.* https://doi.org/10.1002/9780470316566.refs

Peng, K.-L., Hsu, C. H. C., Lin, P. M. C., & Su, M. (2023). Proposing spacetime scale for space tourism economics. *Tourism Economics, 29*(6), 1671–1678. https://doi.org/10.1177/13548166221109666

Peterson, M. J. (1997). Use of analogies to develop outer space law. *International Organization, 51*(2), 245–274. https://doi.org/10.1162/002081897550357

Salman, M., Ali, M., & Bilal, M. (2022). Curvature inheritance symmetry in Ricci flat spacetimes. *Universe, 8*(8), 408. https://doi.org/10.3390/universe8080408

Senadeerage, C. T. (2020). Modelling International Tourism Demand and Implications of Its Volatility for Sri Lanka at a Disaggregate Level.

Song, H., Qiu, R. T. R., & Park, J. (2019). A review of research on tourism demand forecasting: Launching the Annals of Tourism Research Curated Collection on tourism demand forecasting. *Annals of Tourism Research, 75*, 338–362. https://doi.org/10.1016/j.annals.2018.12.001

Song, H., Qiu, R. T. R., & Park, J. (2023). Progress in tourism demand research: Theory and empirics. *Tourism Management, 94*, Article 104655. https://doi.org/10.1016/j.tourman.2022.104655

Tsay, R. S., & Chen, R. (2018). *Nonlinear time series analysis.* John Wiley & Sons.

Uğur, N. G., & Akbıyık, A. (2020). Impacts of COVID-19 on global tourism industry: A cross-regional comparison. *Tourism Management Perspectives, 36*, Article 100744. https://doi.org/10.1016/j.tmp.2020.100744

Wan, L. (2021, November 21). *Commentary: Space tourism by 2024 a growing possibility in China.* Channel News Asia. https://www.channelnewsasia.com/commentary/space-tourism-virgin-galactic-spacex-blue-origin-2319651

Wen, D., Mengxi, H., Li, L., & Zhang, Y. (2022). Forecasting crude oil prices: Do technical indicators need economic constraints? *Quantitative Finance, 22*(8), 1545–1559. https://doi.org/10.1080/14697688.2022.2074305

Williams, A. M., Rodriguez Sanchez, I., & Škokić, V. (2021). Innovation, risk, and uncertainty: A study of tourism entrepreneurs. *Journal of Travel Research, 60*(2), 293–311. https://doi.org/10.1177/0047287519896012

World Travel and Tourism Council (2020). *Japan 2020 annual research: Key highlights.* Available at https://wttc.org/Research/Economic-Impact.

Zhang, H., Song, H., Wen, L., & Liu, C. (2021). Forecasting tourism recovery amid COVID-19. *Annals of Tourism Research, 87*, 103149. https://doi.org/10.1016/j.annals.2021.103149

Chapter 6
Is Effect Size Properly Discussed in Space Tourism Demand Studies?

1 Introduction

While space tourism is still a distant fantasy for most people, it is a future trend that will continue attracting tourists looking for adventure and discovery experiences (Chang, 2017). Since the first manned trip to orbit in 1961, the public has gradually recognized space tourism as a novelty. In particular, attention to space tourism increased dramatically after Dennis Tito's first commercial orbital mission in 2001 (Zhang & Wang, 2022). The national and governmental public bodies controlled the participants and stakeholders in space operations throughout the early stages of developing a space economy, primarily for space exploration and research objectives (Von der Dunk, 2011).

The private sector's involvement increased gradually, speeding up technological advancement and sharply lowering the cost of space flight, ultimately paving the way for the emergence of commercial space tourism (Friel, 2020). Visionary business people like Jeff Bezos, who founded Blue Origin in 2000, Elon Musk, who founded SpaceX in 2002, and Richard Branson, who founded Virgin Galactic in 2004, have been motivated by the hype surrounding the nascent and expanding space experience market to pursue the market. However, it wasn't until the successful launch of SpaceX's Dragon spacecraft in 2020 that the global era of commercial human spaceflight officially arrived. SpaceX officially became the first private company to master the ability to manufacture and launch manned spacecraft (Giachino et al., 2023).

Although a small number of people have traveled in space, the maturity of technology and people's yearning for space will inspire more tourists to explore space. Space tourism is a niche market. There are higher requirements for the financial status of tourists than ordinary tourism. So, space tourism will create profits that normal tourism cannot. Space tourism as a niche sub-sector creates value several times higher than the average industry because of its unique technology and attributes, and this value rapidly makes it attractive. The development of technology and the upgrading

K. Peng et al., *Space Tourism Supply Chain*, Contributions to Management Science,
https://doi.org/10.1007/978-981-95-1080-1_6

of skills will allow the space infrastructure to continue to improve and develop. The launch of rockets will also positively impact launch destinations, leading to the formation of space tourism-related industries in the places where rockets are launched (Friel, 2020). Therefore, exploring the demand for space tourism will have theoretical and practical significance.

This study combed through the existing literature on space tourism to find the literature on the demand for space tourism and then filtered out the articles that have analyzed the effect size of the space tourism demand. We sought to identify effect sizes that have not been adequately discussed in space tourism demand research.

2 Literature Review

Tourism Demand

The study of tourism demand involves several fields, such as economics and psychology. Tourism demand research generally requires drawing on theories and models from various fields of study (Song et al., 2019). They found that four main measures are used in tourism demand-related studies, with causal econometric and non-causal time series models being the most widely used. Due to the increasing importance of combination and hybrid models, more scholars are expected to use them. Song et al. also argue that despite these approaches' theoretical and technological limits, the development of AI-based models is noteworthy in the most recent studies on forecasting tourism demand.

It is worth noting that it has not been long since researchers began applying big data to tourism forecasting. In the future, as technology advances and big data technology matures, big data will play a more significant role in tourism forecasting. However, there are still several difficulties in using big data for tourism demand forecasting. First, new techniques and methods need to be used to deal with both traditional time series and high-frequency big data; second, in addition to mining the data, it is essential to use tourism big data to solve existing problems; and third, how to be able to bring together skilled people with different research directions (Song et al., 2019). Therefore, market demand cannot be quantified or reported as a single number. Instead, it is based on the characteristics of the products offered and the market environment jointly formed by regulators, consumers, and participating businesses.

Space Tourism Demand

It is hardly surprising that there have been numerous studies undertaken on the possible public interest in space tourism. Many of these investigations, conducted in

Japan, the United States, Canada, Germany, and the United Kingdom, were evaluated, and their findings were compared by Crouch (2001). Overall, there was some broad agreement in the results, showing that depending on nationality, gender, and age, more than half a percent of people in these surveys were willing to travel into space. Additionally, according to Crouch, between 10 and 20 percent of respondents said they would be willing to spend their annual salary ($20 million) to travel into space. Laing and Crouch (2004) evaluated Australians' interest in space travel in a comparative study using a survey methodology. Some scholars have also studied the demand for commercial space travel in Australia and found that Australians are about as interested in space tourism as any other country. More than half of them wish they could do a space tour, but their biggest concerns are the price of space tourism, safety, and the design of the vehicle's products (Crouch et al., 2009). Chang (2020) asserts that it is challenging to examine conceptual models of the space tourism sector's possible scale and dependability. Researchers have a little data to apply for a completely new field or other areas to draw from. Everything is unknown, so forecasting a new domain is very difficult.

p-*value and Effect Size*

P-value or significant probabilities have been utilized in place of hypothesis tests to provide more information about the link between the data and the hypothesis than a straightforward reject/do not reject choice (Schervish, 1996). It states that the amount of evidence supporting a hypothesis generated by a scientific observation is called "significance"(Ialongo, 2016).

Ialongo (2016) believed that the null hypothesis is always determinable. It is precisely equal to a particular amount—is a flaw in the framework for conducting hypothesis tests (usually zero). Therefore, expanding the sample would raise the likelihood of obtaining a low p-value, even with a feeble impact. The disadvantage is that we risk misidentifying natural variability or measuring error as a significant effect, especially with biological data. Therefore, to make the results more reliable and accessible for the reader, effect sizes should be provided in addition to the data for p-values. Effect size is the magnitude of the difference between groups.

Effect size is measured using a variety of indices depending on the sort of comparisons being studied. The index can be divided into two broad study categories: group differences in effect sizes and measures of association between variables (Table 1). In the tourism field, association between variables were used most often.

Almost half of the articles reporting analysis of variance (ANOVA) did not include any effect size measures, and only 25% of the survey studies reported additional information about effect sizes in the student t-test analyses, according to a review of studies published in 2009 and 2010 in a reputable psychology journal (Maciej Tomczak, 2014).

Effect sizes are used differently in different content articles, but in the social science field, the more commonly used expression is R^2 or r^2. R^2 values of 0.75, 0.50, or 0.25 for endogenous latent variables in the structural Model can be described as substantial, moderate, or weak, respectively. In regression analysis, the R^2 reflects the proportion of the variance of the dependent variable explained by the independent variable. The stronger the relationship and correlation between X and Y, the R^2 larger. Therefore, pursuing a high R^2 is pursuing a high correlation and a strong linkage between the influences.

To explore the effect size properly discussed in space tourism demand studies or not, this chapter reviewed the literature database of the major tourism fields, find relevant literature on space tourism from 2000 to the present, and screen it for quantitative studies to determine the use of effect sizes.

Tourism Studies Disscussed with Effect Size

In the context of tourism, the term "effect size" is employed to evaluate the magnitude of a phenomenon or intervention within the tourism industry. Effect size is a quantitative measure that quantifies the size of a relationship or difference. It is important to note that effect size extends beyond statistical significance (p-value) and is employed to evaluate practical relevance (Kroes & Finley, 2023). Through the analysis of effect size, researchers are able to assess the impact of various factors on the tourism industry, including the effectiveness of marketing campaigns, changes in tourism policies, or the introduction of new attractions.

Researchers frequently employ effect size to quantify the practical significance of their research outcomes. This involves not only determining statistical significance, but also assessing the practical impact of the results. In the context of tourism research, this aspect is of particular significance, as it facilitates the determination of the practical application value of research outcomes, including policy impact and marketing strategy.

The significance of effect size lies in the potential for even a modest effect to attain statistical significance when a substantial sample size is employed (Liu & Yuan, 2021). Consequently, elucidating the effect size in tourism research is paramount for ascertaining the practical significance of research findings, as opposed to merely maintaining statistical significance. For instance, a study that investigates the impact of social media reviews on hotel bookings would concentrate on effect size. The effect size in this context would denote the magnitude of the relationship between positive reviews and booking rates. An additional example could pertain to the impact of eco-certification on tourist preferences. In this case, effect size would indicate the probability of a tourist choosing an eco-certified hotel over an uncertified one. The significance of effect size can vary depending on the specific context. In some cases, a small effect size may be particularly salient, while in others, it may be less consequential. If a study were to ascertain that a marketing campaign had increased tourist visits, the effect size for that study would demonstrate the magnitude of the increase. While the effect size may be statistically significant, its magnitude may not warrant the campaign's expense if it is minimal (Kroes & Finley, 2023).

In the context of tourism research, as Song et al. (2023) proposed that one of the most frequently employed effect size metrics is Cohen's d, which is utilized to discern differences between groups, such as in the comparison of tourist satisfaction before and after an intervention. Pearson's r, on the other hand, is employed to demonstrate correlations, such as the relationship between tourism and local employment rates. Odds ratios can be used in logistic regression models, for example, to predict the likelihood of someone revisiting a destination based on certain factors. In the field of tourism research, effect size analysis has been employed to examine the impact of digital infrastructure on urban–rural integration. The analysis revealed a significant contribution of digital infrastructure, with an effect value of 0.78. In contrast, government intervention was found to have a negative impact, with an effect value of −0.31 (Tian & Wang, 2025). The threshold regression model reveals the nonlinear impact of trade on tourism, among which the effect value is highest in middle-income countries (Chen & Cheng, 2018). In non-habitual environments, differences in regional landscapes are more likely to reduce tourists' price sensitivity than cultural atmosphere (effect values are 0.85 and 0.62, respectively) (Li et al., 2020). Furthermore, facility completeness (effect value 0.82) and policy effectiveness (0.75) emerge as pivotal factors in shaping tourists' environmental protection behavior (Qin & Hsu, 2022). Likewise, the incorporation of loss frame information has been shown to enhance the carbon compensation selection rate of air passengers by 40% (effect value 0.7) (Ropret et al., 2024). In the context of adventure tourism destination selection, an investigation was conducted to determine the distance desire and its subsequent impact on the destination selection process. The findings reported by Cao et al. (2020) indicated that the effect size of distance desire for game (i.e., the pursuit of novelty) was higher than that of distance desire for fashion, at 0.68, while the effect size of fashion distance desire (primarily referring to social interaction) was 0.45. This observation underscores the significance of game distance desire as a pivotal factor in the determination of preferred adventure tourism destinations and underscores its potential as a critical element in the realm of tourism behavior research. Furthermore, the contribution rate of visual landscape to tourism decision-making (effect value 0.75) surpasses that of auditory stimuli (0.32) (Wu et al., 2023).

Moreover, different fields within the tourism industry may use different benchmarks to evaluate effect sizes. For instance, in economic impact studies, a modest effect size may be sufficient to yield substantial revenue, whereas in behavioral research, a more substantial effect size may be necessary before an intervention is deemed valuable. A meta-analysis in the tourism industry involves the aggregation of effect sizes from multiple studies to draw broader conclusions (Sustacha et al., 2023). This approach facilitates the identification of both commonalities and differences across diverse contexts. To illustrate, a meta-analysis of the effects of cultural experiences on tourist satisfaction might reveal an average effect size, indicating the overall impact of cultural experiences. To enhance the rigor and transparency of tourism research, it is imperative that studies report effect sizes alongside p-values, enabling readers to evaluate the statistical and practical significance of the findings.

This initiative is part of a broader movement towards more transparent and informative statistical reporting in various fields. It is noteworthy that specific disciplines, such as tourism economics, necessitate distinct effect size thresholds compared to those employed in psychology.

When employing effect sizes in analytical endeavors, it is imperative to exercise caution and be mindful of potential pitfalls. For instance, neglecting to consider the research context or sample size can result in a misinterpretation of the effect size. To illustrate this point, consider the example of Cohen's d. Within the field of psychology, 0.5 is typically regarded as a medium effect. However, in the context of tourism economics, this value may bear significantly different implications. Also, relying solely on effect sizes while ignoring their confidence intervals can lead to misleading conclusions. Tourist behavior data are susceptible to memory errors. To enhance the accuracy and reliability of tourist behavior data, it is recommended to combine objective data sources, such as GPS tracking and consumption records. Conversely, overreliance on convenience samples, such as college students, may result in underestimation of effect sizes due to their limited representativeness of the broader population.

In recent years, effect size has emerged as a pivotal instrument in the domain of tourism research, offering researchers a means to assess the efficacy of tourism interventions from the perspective of whether they are "working" to the extent of their observed impact. Through this analysis, researchers can more accurately evaluate the actual value of tourism interventions and promote the transformation of academic results into industrial practice. Effect size assists in quantifying the magnitude of the relationship, evaluating the practical significance of research outcomes, and facilitating the generation of more reliable conclusions concerning tourism phenomena through comparative studies. This tool provides a quantitative basis for understanding tourism behavior and decision-making processes, supporting industry evidence-based decision-making.

Despite the challenges associated with effect size analysis, including issues of contextual interpretation and the establishment of domain-specific benchmarks, future researchers can circumvent potential pitfalls through interdisciplinary integration and technology-driven methods. For instance, functional magnetic resonance imaging (fMRI) was employed to assess the emotional response elicited by destination promotional videos, with the Bali volcano crisis ads demonstrating an amygdala activation effect size of 0.92 (Chen et al., 2020); the positive emotion extension theory was used to reveal that for every 1 unit increase in the richness of sensory experience, tourist loyalty will increase by 0.75 units (Wu & Pearce, 2014); the effect size of using virtual reality to test tourism scene design is 40% higher than that of traditional questionnaires; in addition, sentiment analysis of social media comments (NLP technology) can monitor the effects of tourism policies in real time.

3 Methodology

This study used a rigorous quantitative technique and literature review. In contrast to the conventional narrative review, this form of review offers repeatable and impartial coverage of a specific topic. The knowledge limits can be mapped using this method, and research gaps can be found. This study adopted a systematic five-step literature review method (Pickering & Byrne, 2014). First, decide the goals and objectives of the research; second, establish standards and protocols; third, search and edit the literature; fourth, extract and analyze literature; and finally, synthesize and present related findings. 'Space tourism' and 'Demand' were the keywords to search among the titles in databases. Six academic databases were retrieved to ensure comprehensiveness, including Science Direct, Web of Science, Scopus, Sage, Emerald, EBSCOhost (Hospitality and Tourism Complete) (Zhang & Wang, 2022). Besides, this study only considered academic journal articles published during 2000–2023. Literary periodicals that publish research articles in English were taken into consideration. Not least, this review study focused solely on the social science fields closely related to tourism studies.

424 journal articles were found by keyword search against the databases as of March 2023. These articles were then exported to the Endnote X9 bibliography management program. The authors of the remaining 69 publications carefully studied the abstracts, methods, findings, and conclusions to weed out those that did not fit the selection criteria after eliminating 355 duplicates, irrelevant articles, and articles from natural sciences journals. Even if they may simultaneously explore other kinds and features of space travel, the qualified articles must be based on the social sciences research paradigm and have at least one component relevant to tourism-related issues. 37 studies were ultimately determined to be eligible by the authors. As a result, 13 more research were added, bringing the total number of articles included in the study to 47.

The Vosviewer program version 1.6.19 was used to input these 47 papers and apply them to bibliometric analysis. Vosviewer is a valuable tool for doing systematic literature reviews since it can be used to create and display bibliometric maps and locate word clusters and their reference networks (Waltman et al., 2010).

4 Analysis Results and Discussions

This study not only focuses on the effect size used in the papers but also tries to discover the current research hotspot of space tourism demand. Through screening, only seven articles used quantitative research methods to analyze the demand for space tourism (see Table 1). Notably, Kim et al. (2023) created an extended model of goal-directed behavior (EMGB) and verified it with individual and organizational interventions, comparing space tourism experienced and nonexperience travelers. His paper is the only one that mentioned effect size, but it has not explained the data (see Fig. 1). It seems that the author has not realized the importance of the effect size, which means the effect size was not used properly in this chapter. For the other six papers, except paper 5, which constructed a conceptual model, the other three papers showed the p-value. However, only one paper used the effect size. To make matters worse, other articles do not even mention p-values.

Also, vosviewer's analysis shows that research on the demand for space is very fragmented and disorganized, as indicated in Fig. 2. First, this may be caused by space tourism technology, which is immature and poorly marketed. Secondly, space tourism entails high costs, and a minimal percentage of the world's population can afford to pay. In addition, there are still many unknown risks associated with space tourism that leave the rights and safety of travelers unprotected. However, it can be expected that the development of space tourism will become more and more mature and popular with the advancement of technology. As a result, more and more brave and adventurous tourists will fly to space. Therefore, the study of the demand for space tourism is essential.

Table 1 Paper used quantitative research methods

No	Topic	Author	Years	Methods	p-value	Effet size
1	Modelling consumer choice behaviour in space tourism	Crouch et al.	2008	Quantitative method	Yes	No
2	Space tourism: Research recommendations for the future of the industry and perspectives of potential participants	M. Reddy et al.	2011	Quantitative method	No	No
3	Astrium suborbital spaceplane project: Demand analysis of suborbital space tourism	Le Goff and Moreau	2013	Qualitative and quantitative methods	No	No
4	Social representations of space travel: Modeling the antecedents and outcomes	Tasci et al.	2021	Quantitative method	Yes	No
5	The quest for sustainability in lower orbit: Conceptual models for space tourism	Paladini and Saha	2022	Conceptual models	No	No
6	Is Generation Z ready to fly into the space? The future of tourism is coming	Giachino et al.	2023	Quantitative method	No	No
7	The effects of individual and organizational interventions on space tourism applying EMGB and fsQCA	Kim et al.	2023	Multi-analysis methods	Yes	Yes

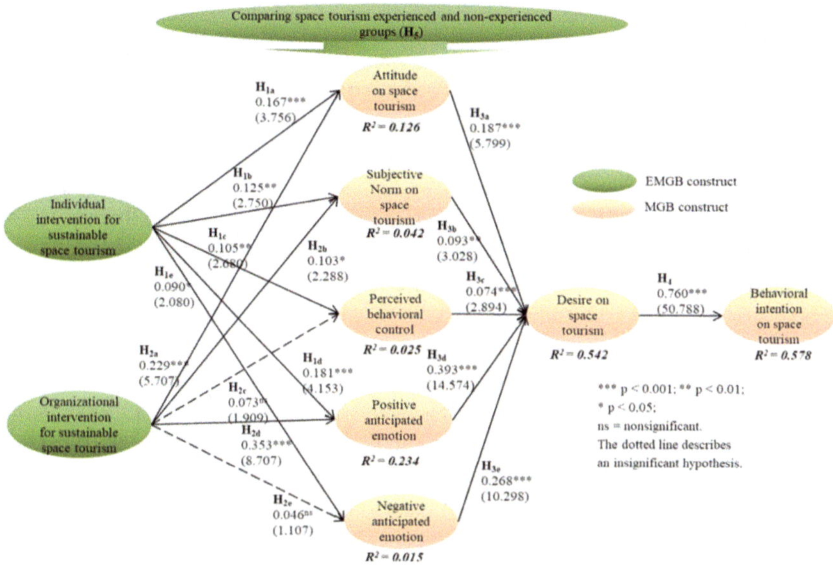

Fig. 1 Models with effect sizes used (Kim et al., 2023)

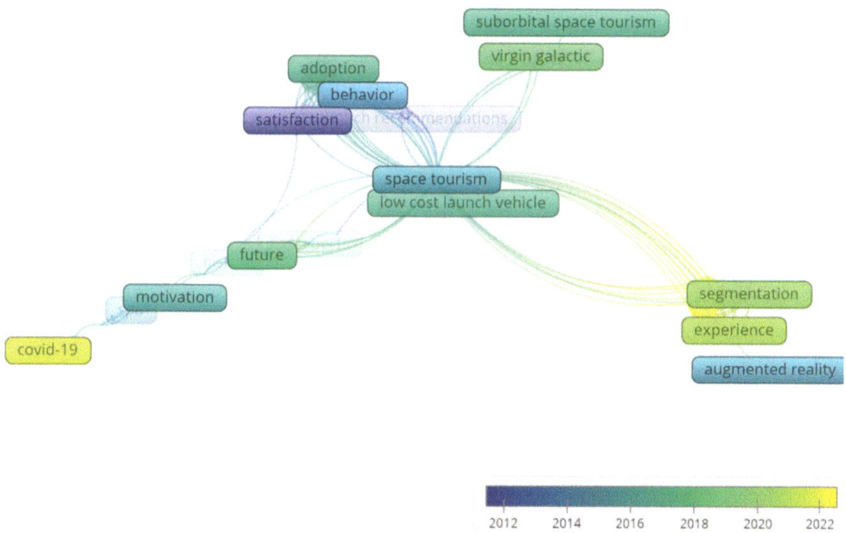

Fig. 2 Vosviewer results

5 Conclusions

Scholars are still having a positive attitude about the development of the space tourism industry. The National Aerospace Laboratory (Japan) showed that the demand for space tourism "apparently has the potential to grow to many times the most optimistic projections for existing commercial space activities" (Collins, 2003). Reddy et al. (2012) stated that space tourism is expected to grow by 18–26% annually from 2020–2030. However, the current academic research on space tourism is still relatively lacking. The real future demand (Crouch et al., 2009) for space tourism needs to be determined. Some companies preparing for space travel have conducted market research, but this is not available for public use.

In future research, more researchers are needed to explore space tourism demand using quantitative methods, such as time series models, econometric models, and artificial intelligence-based models (Song et al., 2019). These models are most used in tourism demand studies. It is also essential to pay attention to applying p-values and effect sizes to improve the reliability and validity of this study. It seems feasible to increase the profile of potential users of space tourism through the use of big data, which can effectively help implement precision marketing strategies and make it easier to understand their personalized needs.

Additionally, it is advised to investigate both the supply and demand sides of the elements impacting consumer behavior for space tourists, including the items for space tourism, the characteristics of space travel, the level of security, etc. (Danov, 2020). This chapter will help future branding and marketing strategies for space tourism.

This study has several limitations. First, this chapter only screened articles related to space tourism and did not focus on any books. Secondly, this chapter only focused on articles that studied space tourism using quantitative methods. Articles that studied space tourism using qualitative and other research methods are also very valuable to study for future research.

References

Antoinette D. A., Kroes Jason R., & Finley (2025). Demystifying omega squared: Practical guidance for effect size in common analysis of variance designs. *Psychological Methods, 30*(4), 866–887. https://doi.org/10.1037/met0000581

Cao, J. J., Zhang, J. H., Wang, C., Sun, J. K., & Chen, M. (2020). Exploring the essence of distance: A framework of distance desire and tourist destination choice. *Acta Geographica Sinica, 75*(4), 860–877. https://doi.org/10.11821/dlxb202004014

Chang, E. Y.-W. (2020). From aviation tourism to suborbital space tourism: A study on passenger screening and business opportunities. *Acta Astronautica, 177*, 410–420. https://doi.org/10.1016/j.actaastro.2020.07.020

Chang, Y. (2017). A preliminary examination of the relationship between consumer attitude towards space travel and the development of innovative space tourism technology. *Current Issues in Tourism, 20*(14), 1431–1453. https://doi.org/10.1080/13683500.2015.1005580

Chen, Q., & Cheng, C. (2018). The Threshold Effect and Interaction Difference between Tourism and Trade Based on the Empirical Evidence of China and 24 Countries. *Tourism Tribune/Lvyou Xuekan, 33*(11), 37–47. https://doi.org/10.3969/j.issn.1002-5006.2018.11.010

Chen, J. Q., Zhang, H. Z., Lu, L., Zhang, H. M., & Xu, Y. C. (2020). Persuasion Effect of Tourism Promotional Videos—A Study of Longitudinal Tracking Experiment Based on Crisis Scenarios. *Tourism Tribune/Lvyou Xuekan, 35*(4), 64–75. https://doi.org/10.19765/j.cnki.1002-5006.2020.04.010

Collins, P. (2003). Tourism market demand and the transportation infrastructure. In *AIAA International Air and Space Symposium and Exposition: The Next 100 Years.* American Institute of Aeronautics and Astronautics.

Crouch, G. I. (2001). The market for space tourism: Early indications. *Journal of Travel Research, 40*(2), 213–219. https://doi.org/10.1177/004728750104000212

Crouch, G. I., Devinney, T. M., Louviere, J. J., & Islam, T. (2009). Modelling consumer choice behaviour in space tourism. *Tourism Management, 30*(3), 441–454. https://doi.org/10.1016/j.tourman.2008.07.003

Danov, D. G. (2020). A Review of Space Tourism Services: Supply and Demand Challenges. *Journal of Tourism Leisure and Hospitality, 2*(1), 29–35.

Friel, M. (2020). Tourism as a driver in the space economy: New products for intrepid travellers. *Current Issues in Tourism, 23*(13), 1581–1586. https://doi.org/10.1080/13683500.2019.1628189

Giachino, C., Pucciarelli, F., Bollani, L., & Bonadonna, A. (2023). Is generation Z ready to fly into the space? The future of tourism is coming. *Futures, 145.* https://doi.org/10.1016/j.futures.2022.103064

Ialongo, C. (2016). Understanding the effect size and its measures. *Biochemia Medica, 26*(2), 150–163. https://doi.org/10.11613/BM.2016.015

Kim, M. J., Hall, C. M., Kwon, O., Sohn, K., & Kim, M. (2023). The effects of individual and organizational interventions on space tourism: Applying EMGB and fsQCA. *Current Issues in Tourism, 26*(9), 1389–1393. https://doi.org/10.1080/13683500.2023.2176744

Kroes, A. D. A., & Finley, J. R. (2023). Demystifying omega squared: Practical guidance for effect size in common analysis of variance designs. *Psychological Methods.* Advance online publication. https://doi.org/10.1037/met0000581

Li, C. X., Feng, H. Y., Lv, Q. Y., & Li, X. Y. (2020). The Impact of Unusual Environmental Differences on Price Perception from the Perspective of Construal Level Theory. *Tourism Tribune/Lvyou Xuekan, 35*(11), 42–53. https://doi.org/10.19765/j.cnki.1002-5006.2020.00.008

Liu, H., & Yuan, K. H. (2021). New measures of effect size in moderation analysis. *Psychological Methods, 26*(6), 680.

Maciej Tomczak, E. T. (2014). *The need to report effect size estimates revisited: An overview of some recommended measures of effect size*

Pickering, C., & Byrne, J. (2014). The benefits of publishing systematic quantitative literature reviews for PhD candidates and other early-career researchers. *Higher Education Research & Development, 33*(3), 534–548. https://doi.org/10.1080/07294360.2013.841651

Qin, Q., & Hsu, C. H. (2022). Urban travelers' pro-environmental behaviors: Composition and role of pro-environmental contextual force. *Tourism Management.* https://doi.org/10.1016/j.tourman.2022.104561

Reddy, M. V., Nica, M., & Wilkes, K. (2012). Space tourism: Research recommendations for the future of the industry and perspectives of potential participants. *Tourism Management, 33*(5), 1093–1102. https://doi.org/10.1016/j.tourman.2011.11.026

Ropret Homar, A., & Knežević Cvelbar, L. (2024). Combatting Climate Change Through Message Framing? A Real Behavior Experiment on Voluntary Carbon Offsetting. *Journal of Travel Research, 63*(4), 883–903. https://doi.org/10.1177/00472875231179400

Schervish, M. J. (1996). P values: What they are and what they are not. *The American Statistician, 50*(3), 203–206. https://doi.org/10.1080/00031305.1996.10474380

Song, H., Qiu, R. T. R., & Park, J. (2019). A review of research on tourism demand forecasting: Launching the Annals of Tourism Research Curated Collection on tourism demand forecasting. *Annals of Tourism Research, 75*, 338–362. https://doi.org/10.1016/j.annals.2018.12.001

Tan J. X., Wang, K. (2025). The impact mechanism of tourismfication on urban-rural integration development in the Wuling Mountain area. Journal of Natural Resources, 40(4), 1048–1067. https://doi.org/10.31497/zrzyxb.20250411

Von der Dunk, F. G. (2011). Space tourism, private spaceflight and the law: Key aspects. *Space Policy, 27*(3), 146–152. https://doi.org/10.1016/j.spacepol.2011.04.015

Waltman, L., Van Eck, N. J., & Noyons, E. C. (2010). A unified approach to mapping and clustering of bibliometric networks. *Journal of informetrics, 4*(4), 629–635.

Wu M., & Pearce, P. L. (2014). Positive Psychology and its Applications in Tourism Studies. *Tourism Tribune/Lvyou Xuekan, 29*(1), 39–46. https://doi.org/10.3969/j.issn.1002-5006.2014.01.004

Wu, X. Y., Liang, X. C., & Song, H. (2023). The tourist sensory experience: A research review and prospects. *Tourism and Hospitality Prospects, 7*(5), 94–119.

Zhang, Y., & Wang, L. (2022). Progress in space tourism studies: A systematic literature review. *Tourism Recreation Research, 47*(4), 372–383. https://doi.org/10.1080/02508281.2020.1857522

Chapter 7
Are Descriptive Statistics of Variables Useful for Modeling Space Tourism Demand?

1 Introduction

Since the first manned orbital flight in 1961, space tourism, as a cutting-edge form of tourism, has gradually entered the public eye and has been dramatically promoted after Dennis Tito's first commercial space flight in 2001 (Zhang & Wang, 2022). The Cambridge Dictionary defines space tourism as "traveling to space for fun and interest, and not as a job" (Danov, 2020). Space tourism, once solely under the purview of governments, has now begun to rapidly develop in the private sector as a form of adventure tourism (Mehran et al., 2023). SpaceX's recent successful launch of the Dragon spacecraft is considered another inspiring milestone for private aviation (Zhang & Wang, 2022). And there is considerable interest in this entirely new service (Danov, 2020). Unlike the political and military drivers of aerospace development in the twentieth century, the internationalization, privatization, and commercialization of aerospace in the new millennium, accompanied by continuous technological advancements such as reusable launch systems and computer science, continue to provide space for tourism. The expanding space industry is driving increased demand for both orbital and suborbital space, with a projected market value of £214.94 billion by 2030, providing a significant opportunity for the industry to evolve and develop further (Mehran et al., 2023). Therefore, the exploration of its needs has been widely valued by researchers.

The current research summarizes the supply and demand of space tourism (Danov, 2020), and makes various predictions and investigations on the motivation (Reddy et al., 2012), attitude (Wang et al., 2021), and decision-making of potential participants (Giachino et al., 2023). However, due to the cutting-edge nature of the field, existing concepts of space tourism are often misleading, people misunderstand the legality and authenticity of commercial space tourism (Olya & Han, 2022), and there are multiple uncertainties in the demand for space tourism. Therefore, the literature on space tourism demand modeling is very limited, and few studies have conducted descriptive statistical analyses of variables in space tourism demand modeling.

K. Peng et al., *Space Tourism Supply Chain*, Contributions to Management Science,
https://doi.org/10.1007/978-981-95-1080-1_7

Using a systematic literature review approach, the present study critically evaluates the demand for space tourism through a review of relevant past research, analysis of the descriptive statistics of the variables, and finally identifies the significance and value of the findings. Specifically, this review has three objectives: (1) to provide an overview of the descriptive statistics of space tourism demand; (2) to discuss the significance of variable descriptive statistics for space tourism demand modeling; (3) to explore the guidance of descriptive statistics on space tourism demand for future research directions. The review of the study will provide a valuable contribution to the field of the demand for space tourism. First, the systematic literature review enriches the body of knowledge on the needs of space tourism. Secondly, the analysis of descriptive statistics can explore the key elements of space tourism demand in detail, which is beneficial for modeling future space tourism demand. Finally, it provides a theoretical basis for future empirical research.

2 Literature Review

Space Tourism Marketing

The concept of "space markets" embraces new start-ups, policies, and threats that profoundly alter existing and new value chains and create new demands (Mehran et al., 2023). Scholars envisioned four stages in the development of space tourism from 2001 to 2040, and Dennis Tito's first flight in 2001 was a pioneering stage. The second stage of the market is limited to the wealthy and adventurous, who are willing to pay a lot of money for this unique venture. This is followed by a phase of rapid growth, with cost reductions due to mature technologies, economies of scale, and competition. Finally, there is a saturation stage (mass market) where growth is driven by economic exaggeration and a new market for young people (Spencer & Rugg, 2004). The development of these four stages is in line with the fundamental law of market development and brings a lot of opportunities to investors. From 2000 to 2018, Silicon Valley space startups attracted about USD 21.8 billion in investments (Denis et al., 2020). Taken together, the market for space travel consists of investors seeking financial opportunities and future passengers wishing to experience space travel (Olya & Han, 2022). Space travel includes several exploratory opportunities, such as new types of accommodation, hotels, and services, that contribute significantly to the evolution of commercial space (Denis et al., 2020) and market demand (Danov, 2020). Space agencies are working with businesses (Olya & Han, 2022) to develop and offer more exciting activities (e.g., space balloons, space burials, use of space hotels) to make space tourism more attractive (Laing & Frost, 2019). Therefore, it is imperative to measure the response of the tourism market (Mehran et al., 2023). The continued success of a company's business model is mainly dependent on the completion of missions that can stimulate return visits and space tourism referral intentions (Danov, 2020).

Space Tourism Demand

The era of commercial space travel officially began in 2020 with the first private spacecraft carrying passengers into space. For anyone brave enough to go into space, space tourism can be considered the travel experience of the future (Giachino et al., 2023). Space tourism can be defined as a part of commercial space travel for recreational purposes (Cohen, 2017), with weightless experience and celestial observation constituting the defining features of space tourism (Chang, 2017). Therefore, the demand for unique, challenging, and exciting experiences drives the demand for space tourism (Reddy et al., 2012). A space tourist is "a person who travels into, to, or through space or celestial bodies for fun and entertainment" (Harrington, 2017). Space tourism companies are actively looking for unique and innovative experiences and are establishing themselves as key players in the market. (Mehran et al., 2023). Therefore, with regard to the modeling of space tourism demand, consideration should be given to various factors from its influencing factors to deal with the concept of space tourists, actual needs, awareness of space tourism, motivation investigation, risk factors, liability and insurance issues, health care and training, legal issues, socio-economic benefits, environmental protection, etc. (Reddy et al., 2012).

According to the researcher, in terms of space tourism demand and market, areas to focus on are cross-geographical and country-specific demand, price fluctuations, existing marketing strategies, and actual growth of the industry (Reddy et al., 2012). With the advancement of technology, the development of space tourism is expected to increase spacecraft reuse, thereby helping to reduce unit transportation costs and meet the needs of more potential tourists. At the same time, the growth of the space tourism economy will increase the demand for other space services such as hotels, hospitals, and entertainment facilities (Wang et al., 2021). In addition, from another macro perspective, space tourism, as an emerging industry, has enormous potential to create new jobs, stimulate economic growth, increase interest in STEM (science, technology, engineering, mathematics) subjects, and improve education (Collins & Autino, 2010). Therefore, using descriptive statistical analysis methods to broadly understand people's attitudes and expectations toward the development of space tourism will be conducive to modeling space tourism demand and promoting the development of space tourism.

Despite the high demand for commercial access to space, fundamental questions are raised regarding the profile of psychographics and situational forces that guide participant responses to space travel and the sustainability of this demand (Mehran et al., 2023). In the space travel industry, consumers may avoid conceptually and practically different services (Crouch et al., 2009), because of the lack of essential elements (e.g., competing products, changes in service features, market share performance) (Olya & Han, 2022).

Challenges of Space Tourism

While the cost of space tourism could potentially decrease, there are still factors that limit its feasibility, mainly related to the inherent risks of space exploration. (Wang et al., 2021). Risk represents a core part of the concerns of space travel participants and is perhaps the most challenging issue facing the global space tourism industry (Reddy et al., 2012). These risks include the dangers of ascent and re-entry, the hazards of the space environment to human physical and mental health, and the possibility of technical failures (Bensoussan, 2010).

Earlier, Ross et al. (2010) suggested that the demand for space travel could dramatically affect the atmosphere, such as exacerbating climate change. Spector (2020) also stated that commercial space companies face operational risk because launches can pose hazards to the area surrounding the launch site, and debris and discarded materials can threaten people, infrastructure, and ecosystems.

As with commercial aviation and all adventure tourism, public perceptions of acceptable risks will be shaped through a social construction process in space tourism. There are so many social, environmental, economic and regulatory risks associated with space travel capabilities that focused issues should be selected for study. Research in every field takes a long time and costs a lot, and there are many hypothetical problems that simply don't happen. We should uphold pragmatism, combine exploration and research, and concentrate resources to solve problems when we encounter or may encounter them.

Space Tourism Studies Disscussed with Desccriptive Analysis

The realm of space tourism has emerged as a dynamic domain within the broader field of tourism research, given its status as an emerging sector. In order to gain a comprehensive understanding of the space tourism market, it is imperative to utilise descriptive statistics as a preliminary step and a fundamental component in modelling the demand for space tourism. This approach has been shown to provide fundamental insights into the relationships, patterns, and characteristics of key factors that influence consumer behavior and market dynamics.

According to the descriptive sstatistics (Mishra et al., 2019), in the context of demand modeling, descriptive statistics play a pivotal role. First, they facilitate the comprehension of the distribution of variables. Descriptive statistics are instrumental in quantifying the central tendency and variability of key variables, including the income level of potential space tourists, age, and frequency of interest in space travel. This quantification is achieved through the mean, median, variance, and skewness. A notable finding is that if the mean income of interested individuals is significantly higher than that of the general population, it suggests that space tourism may initially cater to a wealthier demographic. It also helps to identify risk perceptions. These indicators aid in the identification of outliers, skewed data, or nonlinear relationships that may necessitate transformation, such as log-regularizing income data. Dispersion analysis, encompassing metrics such as range, variance, and standard deviation, quantifies the extent of dispersion within the data. In the context of space tourism, a comprehensive understanding of variation in willingness to pay and interest levels can facilitate the development of effective marketing strategies and pricing models.

Second, associations between variables (e.g., income and willingness to book a flight, safety concerns and purchase intention) can be identified based on the results of the Pearson correlation coefficient. For instance, a study may demonstrate that individuals with high incomes (>$500,000 annually) exhibit a strong propensity for orbital tourism, while the middle-income group demonstrates a preference for suborbital experiences that are more economical and briefer.

Third, descriptive clustering (e.g., k-means) can segment the market based on behaviors such as risk takers (driven by novelty and prestige) and risk averse skeptics (who defer purchases until safety is proven). These market segmentation strategies inform targeted marketing initiatives and demand forecasting. Descriptive statistics establish a benchmark against which the performance of more complex models can be evaluated. For instance, a straightforward mean of past bookings can function as a benchmark for more sophisticated demand forecasting models.

Fourth, descriptive statistics provide a benchmark against which the performance of more complex models can be compared. To illustrate, a rudimentary arithmetic mean of historical reservations can function as a benchmark for a more sophisticated demand forecasting model. Furthermore, the application of descriptive analytics has the capacity to identify and rectify issues, such as missing values, inconsistencies, or biases present in survey data. To illustrate this point, consider a scenario in which a demand model is found to be misrepresenting the elderly population, given that 90% of the survey respondents fall within the 20–30 age demographic. Finally, the utilization of visualization tools based on descriptive statistics, such as histograms, boxplots, and scatterplots, facilitates the comprehension of the relationship between variables. For instance, a scatterplot can illustrate the correlation between income and interest in space travel, thereby indicating whether higher income is associated with higher interest.

A number of studies have sought to analyze the demand for suborbital spaceflight by employing descriptive statistical models. These studies have found that potential space tourists are driven by a desire for unique, challenging experiences, such as viewing the Earth from space and experiencing weightlessness. These so-called "push factors" have been identified as the predominant influence on individuals' preferences regarding destination-specific attributes. Nevertheless, surveys consistently indicate that safety concerns remain a significant barrier, with a study by Reddy et al. (2012) reporting that 70% of respondents in southern England believe space travel is risky, underscoring the necessity for transparency in safety protocols. According to the report by Future Market Insights (March 2025), in 2020, there were virtually no offline bookings, as travel agencies had yet to cater to space tourism. Most interactions were initial inquiries directly with space companies via email or phone. By 2024, the online booking process for space tourism had gained significant momentum, with industry leaders such as SpaceX, Virgin Galactic, and Blue

Origin offering detailed booking systems on their websites. In 2025, online bookings will continue to predominate, with space tourism companies offering customers comprehensive platforms to plan and book space travel. By 2035, online booking will become the predominant booking channel for space tourism, offering tourists a streamlined journey that will be driven by artificial intelligence, from the booking process to post-trip feedback. Orbital Assembly, a prominent player in this sector, aims to operationalize a space hotel by the year 2030, offering luxury amenities such as zero-gravity dining experiences. Projections indicate that this growth will persist at a compound annual growth rate (CAGR) of 10.4% from 2025 to 2035, with the market size anticipated to reach US$5.1 billion by 2035.

The utilization of descriptive analytics empowers researchers to discern the characteristics and behaviors of space tourism visitors and the industry in its entirety. This information can then be used to validate hypotheses about key demand drivers, guiding variable selection in predictive models. Informed decisions regarding market positioning, pricing, and marketing strategies can be made, thereby driving growth in the space tourism industry. The construction of a robust demand model necessitates the integration of descriptive analytics with conjoint analysis, which involves the evaluation of trade-offs such as price versus safety, as well as time series forecasting, which accounts for technological advancements such as reusable rockets. Additionally, scenario testing, which involves the simulation of hypothetical scenarios such as a decrease in ticket prices to $100,000, is employed to assess the impact on demand. It is imperative to acknowledge that while descriptive statistics can discern patterns, they lack the capacity to establish causality. It is imperative to acknowledge the limitations of descriptive statistics and to integrate them with qualitative insights. Moreover, the incorporation of advanced models, such as inferential statistics (e.g., regression analysis) or machine learning (e.g., random forests), is essential for predicting demand under different scenarios.

3 Methodology

This study draws on a five-step literature review procedure, including steps (1) identifying research objectives; (2) formulating review criteria and protocols; (3) searching and refining literature; (4) extracting and analyzing literature; (5) synthesizing and presenting relevant findings.

Based on the research objectives specified in the introduction, we established a review protocol. Initially, we searched the database using the phrase "space tourism" to retrieve publications' titles, abstracts, and keywords. To guarantee the authoritative nature of our data, we conducted a search on the academic database, Web of Science. We screened the titles and abstracts of articles with certain keywords and obtained 133 articles. Furthermore, this chapter exclusively examined academic journal articles published in English language over the past 22 years (2000–2022), coinciding with the emergence of commercial space tourism. The selection period was decided based on research progressions after the inaugural commercial human spaceflight in

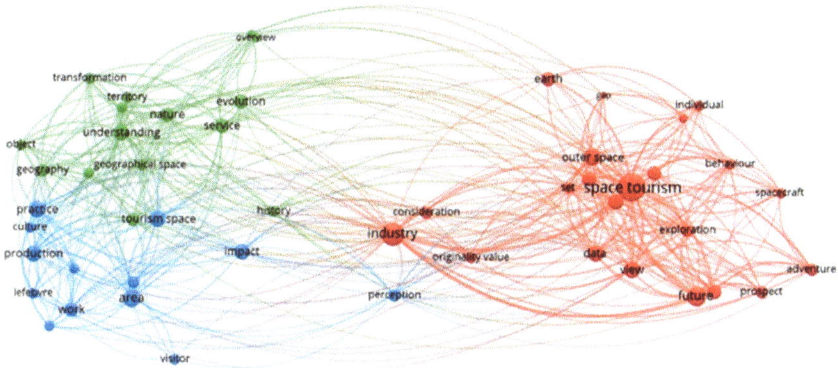

Fig. 1 Space tourism general map

2001. These articles were initially exported to VOSviewer to scrutinize all related keywords. From the Fig. 1, we can see that space tourism is related to the keywords like "behavior,"; "adventure,"; "industry,"; "individual,"; "understanding,"; "nature,"; "service,"; "area,"; "practice." From this map, we can understand the general direction and content of the research and target articles based on the keywords required for demand modeling. We know that space tourism research in recent years has begun to focus on the keyword "individual", from which we can judge that the attitude of potential tourists, that is, individuals, is the focus of future space tourism demand modeling. Then these articles were exported to Endnote for bibliographic management, and after ignoring studies not directly related to market demand and descriptive analysis, articles that did not meet the selection criteria were excluded, and the sample size was reduced to 4 articles.

4 Analysis Results and Discussions

Analyzing the literature with variable descriptive statistical analysis found that (in Table 1) most of the articles were quantitative empirical studies, and some of them had path analysis, which explored many influencing factors. Such as the factor of tourist innovation (Chang, 2017), the interrelationship among motivation, risk perception, and behavioral intention (Olya & Han, 2022), and the moderating effects of motivation, risk factor, and participation, gender, and age (Wang et al., 2021). In addition, the research focuses only on suborbital flights, providing estimates of market size and visitor types (Le Goff & Moreau, 2013).

Table 1 Analysis Results

Sample	Variables	References
British residents	Motivation (Vision of earth from space; Weightlessness; High speed experience; Unusual experience; Scientific contribution)	Reddy et al. (2012)
Generation Z	values; travel choice motivation; space motivation; WTP for specific space tourism experience; interest for specific space tourism experience	Giachino et al. (2023)
Younger and educated	Involvement (Knowledge & Action; General Interest);Motivation (Thrill; Uniqueness; Escape; Challenge; Space experiences; Beauty of the Universe; Science); Perceived risk (Safety and death; Launch and re-entry; Technical failure; Health and fitness condition; Not coming back; Outer space environment; Intra-personal constraints)	Wang et al. (2021)
Potential space tourists from a broader geographical location	Avoidance behavior of space tourism (High perceived risk; High financial cost; Lack of trust; Attitude toward space tourism; Variety seeking; Sign value; Self-esteem; Space experience; Opportunity for exploration; Attractiveness of space)	Olya and Han (2022)
Australians	A specific space tourism option; the price and composition of competing options	Crouch et al. (2009)

Motivations, Attitudes of Potential Participants

Past research has shown that cost, safety, product design, and some demographic and behavioral characteristics can significantly impact attitudes and interest in space tourism (Laing & Crouch, 2004). Afterward, they refer to "mythological constructs" in the culture to examine the underlying motivations for adventurous space tourism (Laing & Crouch, 2011). Qualitative studies include Cater, who based interviews with astronauts to identify underlying drivers (Cater, 2010), and Platt, who constructed a focus group setting to investigate perceptions of space tourism among young Americans (Platt et al., 2020). It has been noticed from past market research reports on expensive adventure activities that people are willing to spend a lot of money to do something that no one or only a few people have done before, such as space travel (Reddy et al., 2012). Desires for space travel include seeking novelty, entertainment, leisure, and intellectual pursuit purposes (Chang & Chern, 2013). Space tourism is geared toward those who pursue their needs at the top of Maslow's pyramid (Maslow, 1943). Wang et al. (2021) stated that motivations for space tourism can also be analyzed based on a push-and-pull framework. Motivations for space tourism include the push factor, such as adventure, exploration, self-esteem, self-actualization, challenge, and thrill-seeking (Laing & Frost, 2019). And two of the most vital pull factors in space travel include novelty and education (Wang et al., 2021). Researchers said space tourism is designed for the few tourists who can accept certain risks and are prepared to do so. The "new" space tourism thus appeals to adventurers and novelty seekers. Perceptions of new adventures and unique experiences, such as weightlessness and seeing Earth from space, drive demand for space travel (Mehran et al., 2023). Thus, two important aspects of the motivations associated with space tourism come into play. First, they are motivated to do something that only a few people have done before, thus becoming the first private astronauts (pioneering aspect); second, the experience of spaceflight, including seeing the earth from space, the feeling of weightlessness and high speed excitement (Reddy et al., 2012).

In a study, the motivations of potential space tourism participants in the United Kingdom were investigated, exploring separately psychosocial motivations for traveling to space (push factors) and specific attributes or characteristics of the product (pull factors). A descriptive statistical analysis of the motivational variables is presented. The study shows the number of samples, mean, and standard deviation, respectively, and the five motivation variables are divided into "seeing the earth from space," "unusual experience," "weightlessness," "high-speed experience," and "scientific contribution." The average value analysis shows that "seeing the earth from space" is the most critical motivation, and "scientific contribution" is the least essential motivation. Through descriptive analysis, it is possible to understand the primary motivations of the research objects from a more comprehensive perspective and to demonstrate the overall choices of the participants, which reflects the significance and value of descriptive statistics. Motivation can promote demand, and as an antecedent variable of demand, it will be beneficial to modeling space tourism demand in the future (Reddy et al., 2012).

When the experience of space tourism is opened to the public, young people may be the generation that will have the opportunity to feel the experience. The future of space tourism depends on how young generations perceive it, and it will be crucial to the growth of commercial space tourism (Wang et al., 2021). In the study, the gender, values, travel choice motivation, space tourism motivation, specific space tourism experience to willingness to pay, and interest in particular space tourism among potential space tourism participants of Generation Z were investigated. Descriptive statistical analysis was performed on the above variables, showing the frequency of the variables, from which it can be clearly found that the characteristics of the respondents are generally fearful of the unknown and oriented towards human well-being. In addition, the respondents are interested in the space experience. Still, their interests are mainly focused on aspects other than the space experience, such as the duration and departure location of the trip, the type of aircraft used, the nationality of the travel provider, the travel risks, etc., and they are willing to pay very little or not for the spaceflight experience. However, most respondents agree that orbital and suborbital spaceflight is an exciting experience, and men's interest and willingness to pay are generally higher than women's. Descriptive statistical analysis can quickly identify the general tendency of respondents and help make trend analysis (Giachino et al., 2023).

The study conducted by Wang et al. (2021) investigated knowledge and action factors as well as general interest in young people's participation in space tourism.

The findings demonstrate that the 35+ age group displayed greater knowledge and higher interest in comparison to the two younger groups. The data showed that the 35+ age group had more knowledge and higher interest than the two younger groups. Additionally, another survey explored the motivations of potential young space tourism participants, with the push factors being thrill, uniqueness, escape, and challenge, and the pull factors being the space experience, beauty of the universe, and science. A descriptive analysis of variables showing their means and standard deviations revealed that the motivations of uniqueness and the beauty of the universe factors were the most important to the survey respondents.

Constraints to Potential Participants

As space tourism has been hampered by technical and financial setbacks, many predictions about its imminent debut have proven premature (Cohen & Spector, 2020). Numerous past studies have examined potential participants' concerns about space tourism because it involves complex processes. Potential tourists may avoid these trips at any stage due to various risks and costs (Olya & Han, 2022). Reddy et al. (2012) found that risk is potentially tourists' primary reason for unwillingness to travel into space (34% of total responses). The risk factors, including safety risks and psychological risks, impact space tourists' behavior significantly and adversely. An earlier study of Australian residents' attitudes toward space tourism found that, besides those who stated they were not interested in space travel, other reasons cited included danger, cost, physical discomfort, and lack of technical knowledge. Angle is skeptical that space tourism will become a reality. The sensory appeal of space tourism is mainly limited to the eyes and vestibule (the sense of balance affected by weightlessness). Other senses remain largely idle, especially hearing and touch (Cohen & Spector, 2020). Notably, space travel will involve long transit times and severe health threats, such as decreased bone and muscle density (caused by weightlessness) and cancer (caused by radiation). Therefore, people may prefer to imagine space travel in virtual reality. Due to the small sample size of those who have traveled in space, and the many new changes in technology over time, the risk of space flight has not been objectively estimated (Wang et al., 2021), so there are still many opportunities in the future to alleviate people's risk concerns.

Natural environmental issues in space tourism research mainly stem from sustainable tourism studies. As in the case of the growing debates in the aviation industry, the critiques for mass space tourism have equally been put forward. Many researchers have proposed that the presence of tourists in space would cause severe earth and space environmental problems, such as the overwhelming carbon and particulate matter emission, leading to air pollution, climate change, and environmental deterioration on the earth, as well as the debris, garbage, and pollution in the outer space.

As space tourism matures, it may provide customers with entirely new experiences in the future, greatly expanding the reach of tourism, but its development also brings new constraints (Cohen & Spector, 2020). Peeters (2018) has suggested that space tourism for recreational purposes, even if limited to the wealthy elite, is highly harmful to Spaceship Earth. Space tourism can be an environmental disaster for reasons including emitting carbon at high altitudes, disrupting celestial bodies' environments, altering them to mimic Earth (terraforming), polluting outer space with garbage, and using space to promote political hegemony and Potential militarization. Therefore, studies have shown that sending tourists to other celestial bodies for short trips may lead to the depletion of Earth's energy sources (Cohen & Spector, 2020), which may have adverse social and environmental impacts, and the booming suborbital industry will significantly exacerbate climate change. Variation (Ross et al., 2010), the space tourism industry is not yet based on a comprehensive regulatory regime. Therefore, there are still various obstacles for potential participants.

The study by Wang et al. (2021) investigated the perceived risk of young people participating in space tourism, and the descriptive analysis of the variables showed frequency. The data showed that the youngest male group had the lowest perceived risk of space travel, and the youngest female group had the highest perception of space travel risk. In addition, perceived risks to space tourism were grouped into eight categories. It is clear from the data that the largest category of perceived risk involves general concerns about personal safety and death (21%). Therefore, the study concludes that future orbital space tourists can be described as men with high-risk tolerance, genuinely interested in space travel-related issues, and want to experience the thrill of space travel.

Due to the uncertainty associated with space tourism, potential tourists perceive the risk to be high, and high perceived risk, high financial costs, and lack of trust increase avoidance of space tourism. The studied the avoidance behavior of potential participants, conducting questionnaires from three perspectives. Descriptive statistical analysis of variables included mean, standard deviation, and deviation. Skewness and kurtosis were used to test the normality of the data. The data indicates that despite concerns over personal risk and safety, lack of trust, and cost, there is significant and ongoing interest in space tourism. Moreover, the positive sentiment towards space tourism outweighs the negative sentiment; thus, the industry represents a worthy future investment (Olya & Han, 2022).

5 Conclusions

The existing space tourism literature is still in its infancy, and there are few current studies, most of which are conceptual papers. Some empirical studies also focus on market research (Zhang & Wang, 2022). There is a lack of scientific research on general and broad concepts of space tourism regarding marketing and psychology (Mehran et al., 2023) with substantial opportunities for academic inquiries. Despite the rising trend of publications of space tourism studies in the past two decades, space tourism research is still in an infant stage, particularly evidenced by the limited number of extant empirical studies. With the progress of commercialization and technological advancements, both space tourism practices and the amount of empirical data are expected to grow swiftly, further enhancing the necessity and feasibility of space tourism research. Therefore, it could be reasonably predicted that, in the following decades, more space tourism studies will be conducted, more special issues for space tourism will turn up, and a new journal for space tourism could also be established.

Meanwhile, space tourism research per se is highly interdisciplinary and complex. As indicated from the thematic analysis results, space tourism issues are related to diverse tourism research domains and have been investigated by scholars from diverse realms. Consequently, this interdisciplinary and complex nature leads to various theories and methods being adopted in space tourism research.

Aside from the conceptual research, empirical studies are insufficient. As tourism research normally has either the economic planning-management or ecology-anthropology-sociology orientation, the suggestions are categorised into two directions. The first direction is harnessing the increasingly available primary data for market research. It is suggested to investigate both push and pull factors influencing the space tourist's consumer behaviors, such as the space tourism products, features of space trips, security level, etc. on the supply side and various demographic, psychographic, and behavioral factors on the demand side. These studies will further benefit future space tourism branding and marketing practices.

To sum up, the descriptive statistical analysis of variables is generally combined with quantitative research methods, and it is the analysis and presentation of the results of structured questionnaire surveys. Through the summary of the space tourism literature on the descriptive statistical analysis of variables, it is found that most of the literature is used to measure and analyze the demographic characteristics and motivation of potential participants. Through the analysis of these documents, we can deeply understand the current situation of space tourism in terms of opportunities and challenges. In terms of challenges, future participation in high-risk tourism activities can be improved through information-rich planning and risk communication programs (Olya & Al-ansi, 2018), providing effective communication and educating space tourists to address functional and safety risks and lack of Trust (Olya & Han, 2022). In addition, marketers can strategically develop personalized marketing communications strategies for those with high marker values and positive attitudes towards space tourism. Most space tourism research has focused on the perspective of potential tourists. To better visualize space tourism through descriptive statistical analysis, we can focus on the understanding and investigation of service providers (including technical support staff working for aerospace companies), and study space tourism from another perspective. Descriptive statistical analyzes of different samples allow for a finer-grained profile of possible orbital space tourists, increasing the broad understanding of people's desire to participate in space tourism (Wang et al., 2021). In the future, if one of these samples conducts space travel, analysing it will provide invaluable data for the civilianisation of space tourism.

References

Bensoussan, D. (2010). Space tourism risks: A space insurance perspective. *Acta Astronautica, 66*(11–12), 1633–1638. https://doi.org/10.1016/j.actaastro.2010.01.009

Cater, C. I. (2010). Steps to space; opportunities for astrotourism. *Tourism Management, 31*(6), 838–845. https://doi.org/10.1016/j.tourman.2009.09.001

Chang, Y.-W., & Chern, J.-S. (2013). From spirit of St. Louis to SpaceShipTwo: A study on the challenge of future space tourism development. *Journal of Aeronautics, Astronautics and Aviation. Series A, 45*(2), 135–154.

Chang, Y. W. (2017). A preliminary examination of the relationship between consumer attitude towards space travel and the development of innovative space tourism technology. *Current Issues in Tourism, 20*(14), 1431–1453. https://doi.org/10.1080/13683500.2015.1005580

Cohen, E. (2017). The paradoxes of space tourism. *Tourism Recreation Research, 42*(1), 22–31. https://doi.org/10.1080/02508281.2016.1239331

Cohen, E., & Spector, S. (2020). Space tourism-past to future: A perspective article. *Tourism Review, 75*(1), 136–139. https://doi.org/10.1108/tr-03-2019-0083

Collins, P., & Autino, A. (2010). What the growth of a space tourism industry could contribute to employment, economic growth, environmental protection, education, culture and world peace. *Acta Astronautica, 66*(11–12), 1553–1562. https://doi.org/10.1016/j.actaastro.2009.09.012

Crouch, G. I., Devinney, T. M., Louviere, J. J., & Islam, T. (2009). Modelling consumer choice behaviour in space tourism. *Tourism Management, 30*(3), 441–454. https://doi.org/10.1016/j.tourman.2008.07.003

Danov, D. G. (2020). A review of space tourism services: Supply and demand challenges. *Journal of Tourism Leisure and Hospitality, 2*(1), 29–35. https://dergipark.org.tr/en/pub/toleho/issue/53785/739977

Denis, G., Alary, D., Pasco, X., Pisot, N., Texier, D., & Toulza, S. (2020). From new space to big space: How commercial space dream is becoming a reality. *Acta Astronautica, 166*, 431–443. https://doi.org/10.1016/j.actaastro.2019.08.031

Giachino, C., Pucciarelli, F., Bollani, L., & Bonadonna, A. (2023). Is generation Z ready to fly into the space? The future of tourism is coming. *Futures, 145*, 103064–103076.

Harrington, A. J. (2017). Legal and regulatory challenges to leveraging insurance for commercial space. *Journal of Space Law 41*, 29.

Laing, J., & Frost, W. (2019). Exploring motivations of potential space tourists. In *Space tourism* (Vol. 25, pp. 141–161). Emerald Publishing Limited. https://doi.org/10.1108/S1571-504320190000025008

Laing, J. H., & Crouch, G. I. (2004). Australian public interest in space tourism and a cross-cultural comparison. *Journal of Tourism Studies, 15*(2), 26–36.

Laing, J. H., & Crouch, G. I. (2011). Frontier tourism: Retracing mythic journeys. *Annals of Tourism Research, 38*(4), 1516–1534. https://doi.org/10.1016/j.annals.2011.02.003

Le Goff, T., & Moreau, A. (2013). Astrium suborbital spaceplane project: Demand analysis of suborbital space tourism. *Acta Astronautica, 92*(2), 144–149. https://doi.org/10.1016/j.actaastro.2013.03.025

Maslow, A. H. (1943). A theory of human motivation. *Psychological Review, 50*(4), 370.

Mehran, J., Olya, H., & Han, H. S. (2023). Psychology of space tourism marketing, technology, and sustainable development: From a literature review to an integrative framework. *Psychology & Marketing*. https://doi.org/10.1002/mar.21795

Mishra, P., Pandey, C. M., Singh, U., Gupta, A., Sahu, C., & Keshri, A. (2019). Descriptive statistics and normality tests for statistical data. *Annals of cardiac anaesthesia, 22*(1), 67–72. https://doi.org/10.4103/aca.ACA_157_18

Olya, H., & Han, H. (2022). Emerging space tourism business: Uncovering customer avoidance responses and behaviours. *Journal of Vacation Marketing, Article, 13567667221101408*. https://doi.org/10.1177/13567667221101408

Olya, H. G., & Al-ansi, A. (2018). Risk assessment of halal products and services: Implication for tourism industry. *Tourism Management, 65*, 279–291. https://doi.org/10.1016/j.tourman.2017.10.015

Peeters, P. (2018). Why space tourism will not be part of sustainable tourism. *Tourism Recreation Research, 43*(4), 540–543. https://doi.org/10.1080/02508281.2018.1511942

Platt, C. A., Jason, M., & Sullivan, C. J. (2020). Public perceptions of private space initiatives: How young adults view the SpaceX plan to colonize Mars. *Space Policy, 51*, Article 101358.

Reddy, M. V., Nica, M., & Wilkes, K. (2012). Space tourism: Research recommendations for the future of the industry and perspectives of potential participants. *Tourism Management, 33*(5), 1093–1102. https://doi.org/10.1016/j.tourman.2011.11.026

Ross, M., Mills, M., & Toohey, D. (2010). Potential climate impact of black carbon emitted by rockets. *Geophysical Research Letters, 37*(24). https://doi.org/10.1029/2010GL044548

Spector, S. (2020). Delineating acceptable risk in the space tourism industry [Article]. *Tourism Recreation Research, 45*(4), 500–510. https://doi.org/10.1080/02508281.2020.1747798

Spencer, J., & Rugg, K. L. (2004). *Space Tourism: Do you want to go?* Collectors Guide Pub.

Wang, L., Stepchenkova, S., & Kirilenko, A. P. (2021). Will the present younger adults become future orbital space tourists? *Tourism Recreation Research, 46*(1), 109–123.

Zhang, Y., & Wang, L. (2022). Progress in space tourism studies: A systematic literature review. *Tourism Recreation Research, 47*(4), 372–383.

Chapter 8
Which Behavioral Economics Concepts Have Been Applied in Space Tourism Demand Analyses?

1 Introduction

Tourists display constrained rationality when they employ heuristics to simplify complex judgments or make decisions that aren't exclusively driven by their interests. Behavioral economics, which goes beyond the classical economics' sometimes challenged assumptions about rational economic conduct, provides the theoretical underpinnings and methodology for research in tourism and hospitality (Li et al., 2022). It is evident that there is a growing usage of behavioural economic and behavioural finance concepts in the analysis of tourism demand, as attested by the studies of Song et al. (2023). Nicolau's research (2008, as cited in Song et al., 2023) examined the effects of reference dependence, loss aversion, and diminishing sensitivity to tourism prices within the Spanish tourism industry, and is widely acknowledged for introducing behavioural economic principles to the tourism and hospitality sectors. Fundamentally, behavioral finance aims to provide a human perspective on the what, why, and how of finance and investing (Ricciardi & Simon, 2000). Although behavioral economics does not replace traditional economics, it usefully extends the rational choice and equilibrium models (Ho et al., 2006). Since 2016, a multitude of studies have researched the implementation of behavioural economics and behavioural finance in the travel and hospitality industries (Li et al., 2022; Song et al., 2023). These studies examine human behaviour, including organized behaviour, from the perspectives of behavioural psychology and economics Few studies have provided experimental data on fundamental concepts, such as complementarity versus substitution, elastic and inelastic demand, and open and closed economies (Hursh, 1984).

The primary purpose of this article is to sort out the concepts related to behavioral economy and behavioral finance and their application in tourism. In addition, on this basis, combined with space tourism demand analysis, this study explores what behavioral economic concepts can be applied to space tourism research.

K. Peng et al., *Space Tourism Supply Chain*, Contributions to Management Science,
https://doi.org/10.1007/978-981-95-1080-1_8

2 Literature Review

Behavioral Economic and Tourism Demand

Behavioral Economic and Behavioral Finance

Economic theory typically presupposes that people approach complex problems in an economist's manner (Thaler & Benartzi, 2004). "Behavioral economics" promises to bring psychology and economics back together by increasing the realism of the psychological assumptions that underlie economic theory. Better predictions of economic behavior and more effective policy recommendations should result from reunification (Camerer, 1999). According to Hursh (1984), economic theory extends beyond mere comparison to behavioral psychology; it constitutes a field of behavior research in its own right, albeit one focused on highly organized patterns of human behaviour. Unlike behavioral psychology, economic theory lacks the same strong empirical foundation with regards to individual subject-controlled experiments.

Otherwise, the utility of economic concepts for behavioral psychology depends on (1) their empirical validity when tested in a lab with individual people and (2) their comparability to other well-established behavioral notions. Similarly, behavioral finance aims to clarify and improve knowledge of investors' thought processes (in Fig. 1), particularly the emotional processes involved and the extent to which they impact choice (Ricciardi & Simon, 2000).

In Table 1, Camerer (1999) summarized the key ideas in behavioral economics. As an illustration, the Expected Utility Theory (EU), which forms the foundation for theories of asset pricing, insurance purchasing, corporate structure, and private decisions like education investments, contends that people evaluate risky bets by weighing the usefulness of an occurrence.

Field phenomena have also been predicted and explained using behavioral economics. Moreover, behavioral economics would offer a more sensible and practical framework for formulating economic policy. Policies are not essential to assist

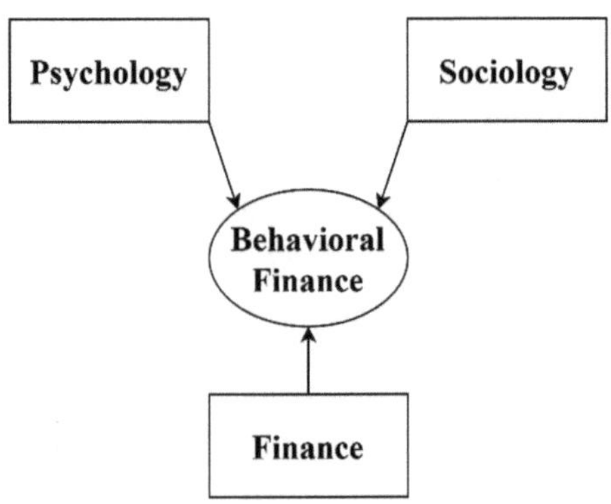

Fig. 1 Definition of behavioral finance (Ricciardi & Simon, 2000)

Table 1 Behavioral economic concepts

Concept	Definition	Note
Expected Utility Theory	The anticipated utility (EU) theory states that people assess risky wagers by evaluating the benefit of a probable outcome. EU assumes that people consider the outcomes in the context of their wealth. If two wagers have the same probability of a specific outcome, that outcome is disregarded when deciding which bet to place	In the EU, a potential outcome is weighed according to its probability. In contrast, people are presumptively going to consider a possible consequence using a "decision weight," which is a nonlinear transformation of the possibility of the outcome. EU is the basis for theories relating to asset pricing, insurance purchases, corporate structures, and private choices like educational investments
Exponential Discounting	People must, in some way, assess these potential costs and advantages to make decisions. Exponential discounting is a presumption in the rational appraisal of possible outcomes	Exponential discounting assumes that the only factor affecting how two payments are compared is the time that has passed since the two payments were made
Social Utilities	Social utility assumes that people want reciprocity or oppose allocations in which they earn less than others to explain these tendencies. These theories are frugally priced, can account for a surprising variety of distinct experiments (like frequent collaboration), and can even foresee specific novel patterns	In "ultimatum games," one player gives $10 to another player while keeping the remaining $5 for herself. The answering player has two options: accept the offer and provide them with both nothing or reject it and give them both nothing. Players who want to maximize their wealth will take anything. Therefore, the first player should make very little of an offer. Unexpectedly for economists, players often offer between $4 and $10 in studies across various nations, some with enormous stakes, and low offers of less than $2 are rejected 50% of the time
Equilibrium	In a market, equilibrium means that supply and demand are equal; in a game of strategy, it means that everyone is making the best choices they can (given that others are too)	

Source Revised from Camerer (1999)

reasonable individuals because they rarely make mistakes. So, relaxing rationality assumptions allows for logical discussion on how people can be assisted. As behavioral economics is incorporated into mainstream economic theory and is applied to other fields, it will no longer be a distinguishing moniker for an approach (Camerer, 1999).

Concepts of Behavioral Finance

The classical and neoclassical financial theory, which now dominates financial analysis, has the potential to be a helpful complement to behavioral finance. It considers psychological variables a significant input to economic analyses and decisions and is picking up steam in academic studies (Fromlet, 2001).

Overconfidence is the tendency to overestimate the likelihood of a series of events (Mahajan, 1992). In operational terms, this can be determined by assessing whether a particular probability exceeds the proportion of correct assessments for all evaluations given that probability (Ricciardi & Simon, 2000). In the field of behavioral finance, the concept of cognitive dissonance plays a prominent role. This is especially relevant within the framework of the Theory of "Financial Cognitive Dissonance". Festinger's theory of cognitive dissonance posits that people experience internal tension and anxiety when they are confronted with conflicting beliefs, a notion supported by Ricciardi and Simon (2000). According to the regret theory, individuals assess their predicted reactions to a future event or situation (such as a loss of $1,000 from selling IBM stock). Regret is described by Bell (1982) as the emotion that arises from comparing an outcome or concern to a previous decision (Ricciardi & Simon, 2000). Table 2 presents the relevant ideas in the behavioral finance field. This checklist encourages using behavioral finance principles to explain atypical economic behavior.

Table 2 The behavioral finance checklist (Ricciardi & Simon, 2000)

The Behavioral Finance Checklist	
Anchoring	Panics
Financial Psychology	Disposition Effect
Cascades	Loss Aversion
Chaos Theory	Prospect Theory
Cognitive Bias	Regret Theory
Cognitive Dissonance	Groupthink Theory
Cognitive Errors	Anomalies
Contrarian Investing	Market Inefficiency
Crashes	Behavioral Economics
Fear	Overreaction
Greed	Under-reaction
Herd Behavior	Overconfidence
Framing	Mental Accounting
Hindsight Bias	Irrational Behavior
Preferences	Economic Psychology
Fads	Risk Perception
Heuristics	Gender Bias
Manias	Irrational Exuberance

Behavioral Economic in Tourism Demand Analyze

Behavioral economics (BE) studies how psychological, cognitive, emotional, cultural, and social aspects affect the financial choices that people and organizations make. Based on economics and psychology, BE is used to comprehend a range of anomalies that are challenging to explain using conventional economic models (Li et al., 2022; Wilkinson & Klaes, 2017). BE has been used to analyze organizational behavior (Camerer & Malmendier, 2007), marketing (Ho et al., 2006), educational investment and results (Koch et al., 2015), and reproductive health issues (Action, 2015). Yet, not enough thought has been given to how BE may account for tourist behavior (Li et al., 2022; Tan et al., 2018). Given the competitiveness of the tourism economy, applying BE to studies in tourism and hospitality may help service providers create more successful business strategies (such as pricing and sales), and destination management organizations promote their destinations more successfully (Li et al., 2022; Neto et al., 2020).

Consumers often make unconventional selections due to internal emotional changes or surrounding influences. Consumers might not have the time or resources to gather all the data required to evaluate their service. Framing, restricted attention, menu effects, persuasion, social pressure, and emotions are all components of non-standard decision-making. Emotions are closely tied to tourist experiences and significantly impact how tourists make decisions (Rind & Bordia, 1996). Hanks et al. (2016) explored the use of superstition signals in advertisements and found that individuals who feel well hold higher expectations for a team's performance, exhibit a more positive outlook, and are more likely to express a strong intention to purchase. Other manifestations of irrational decision-making are known as the "limited attention effect", whereby information is given disproportionate weight or ignored altogether as a consequence of insufficient consideration of a particular choice. To put it another way, restricted attention occurs when a person bases their decisions just on a subset of the information that is readily accessible owing to ignorance of, or a lack of clarity surrounding, shipping costs in eBay auctions or ambiguous taxes (Gabaix & Laibson, 2006). One can discern the necessity of social pressure and persuasion when a football referee adds extra time to a match following a one-goal loss for the home teamNon-standard beliefs are deviations from the economic norm in a consumer's perception caused by three things: projection bias, the law of tiny numbers, and overconfidence (DellaVigna, 2009). These elements could result from underestimates, overestimation, or a lack of knowledge. Overconfident people tend to overestimate their abilities, including their capacity to finish a task or reach a goal. The overconfidence above is primarily blamed on faulty estimation or inadequate information gathering, such as CEOs underestimating their capacity to manage their businesses (Camerer, 1999).

Understanding and enhancing competitiveness in the visitor economy require applying of BE to research in the areas of tourism and hospitality industry. Li et al. (2022) conducted a systematic review of the application of behavioural economics in the tourism industry. The gathered literature was summarised, and the concept of the application of behavioural economics in the field of tourism was illustrated as

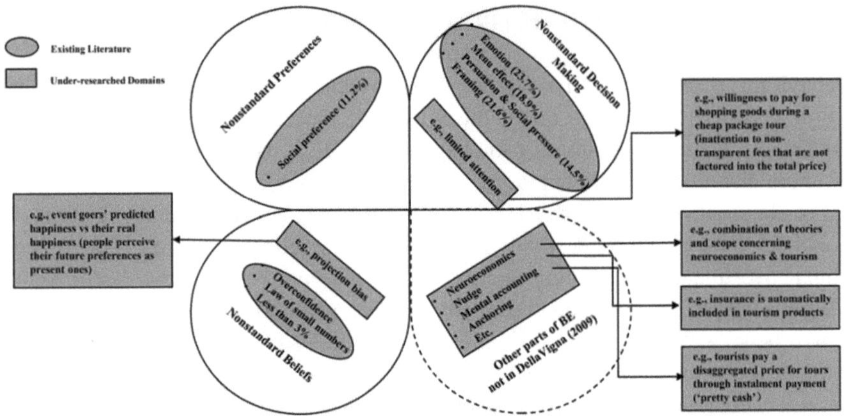

Fig. 2 Application of behavioural economics in tourism (Li et al., 2022)

shown in Fig. 3. This figure amply illustrates the current interplay between behavioral economy and tourism. Based on Fig. 2, it aids in our comprehension of BE ideas and helps us identify which concepts can be applied to describe and analyze real-world phenomena. To help tourism businesses and destinations become more competitive in this fiercely competitive market climate, creative views as well as frameworks for understanding the behavior and decision-making processes of major tourism stakeholders are essential.

Figure 3 illustrates how Park et al. (2018) investigated the links between reciprocity, tour satisfaction, and overall perceptions of the city utilizing social preferences. Risk preferences are utilized to investigate how people's attitudes about risk influence their behavior. Prospect theory, comprising of five components and often employed in risk-taking research, includes risk avoidance over gain-seeking, contemporaneous preference for insurance, gambling, and limited framing, which are among the risk-taking behaviors that should be prioritized (DellaVigna, 2009; Li et al., 2022). Hernandez et al. (2019) employed loss aversion and reference dependence to examine how to wait times at theme parks impact guests' willingness to pay (WTP) for an express pass. In early research, behavioral economics was utilized as an exploratory framework to look at phenomena connected to tourist demand that the ideas of economics or marketing could not fully explain. The application of behavioural economics to the tourism industry therefore has much room for growMoreover, behavioral economics is currently employed to describe the tourism phenomena, primarily from the industry's macro level and rarely for a specific demographic or type of tourism. Hence, a future study will focus on applying behavioral economics to the demand investigation of various forms of tourism. For instance, little research has been done on space tourism. Figure 3 shows how Park et al. (2018) used social preferences to examine how reciprocity, tour satisfaction, and general impressions of the city are related. Risk preferences are utilized to investigate how people's attitudes about risk influence their behavior.

Space Tourism and Behavioral Economic

Demand Analyze in Space Tourism

The number of tourists who travel to witness space shuttle launches, observe a complete solar eclipse, or take part in cosmonaut training highlights the interest in terrestrial space tourism (Bunghez, 2015). A 1997 survey by the Space Transport Association (STA) revealed that three million Americans were ready to spend US $ 100,000, and 40 million would contemplate embarking on a space shuttle journey. (Bunghez, 2015). The appeal of space tourism, or orbital tourism, as a potential target market for businesses, will be critically dependent on the size of the market for space tourism services in the future at various price points (Collins et al., 1994). The exceedingly wealthy are currently the only ones who can travel to space. The market with the most potential for introducing space travel to the general public is suborbital flight, which Virgin Galactic controls. Bunghez's (2015) research demonstrates notable public interest in this activity. Yet, before contemplating providing this service, business owners should meticulously review crucial factors from both consumer and business standpoints. Additionally, he uncovered substantial demand for space tourism amongst consumers. This study considers several factors, including the correlation between customer demand, cost, and essential elements of the space tourism experience that can convince consumers to invest significantly.

In the concise history of space tourism, it is clear that the majority of travellers have embarked on only one journey. However, individuals have already embarked on several space expeditions. Nevertheless, factors beyond cost play a significant role in the advancement of space tourism services, and therefore, such aspects should receive meticulous consideration (Danov, 2020). In addition to pricing, behavioral and psychological aspects, and motivations likely affect the desire for space tourism. To better understand the desire for space tourism, behavioral economists started to apply ideas from behavioral economy and behavioral finance. The preferences of tourists are frequently observed to be inconsistent, maybe due to the lack of complete information or heuristic decision-making (Tan et al., 2018). Many researchers and governmental and commercial entities conducted market surveys in the UK, Japan, Canada, and the US to better know and examine the target market's requirements and preferences (Bunghez, 2015). Yet, evaluating research findings in light of potential changes brought on by factors such as price, travel safety, and product design offer is challenging.

Behavioral Economic in Space Tourism

Similar to how people in the nineteenth century gazed to the skies and imagined a time when powered flight would be possible, people in the twenty-first century also dream of space travel and exciting adventures in the far reaches of the cosmos. Mass space tourism seems to be beyond the capabilities of current technology within these constraints. But, as history has often shown, new technologies are likely to develop to produce previously unimagined answers to the issues of the modern period. Only wealthy people are prepared to pay since earlier studies have shown that those who can afford an adventure lasting around two weeks aboard the International Space Station can do so for 20 to 40 million dollars (Bunghez, 2015). So, much research is being done to examine the requirements and drivers for space tourism. Given the cost, what would motivate the wealthy to spend so much money on a space journey? Is this due to rich people having more money than they can spend? They don't fear dying, do they? Perhaps this is not the most economically sound way to approach the issue. Several studies have covered space travel's macro and micro financial and economic aspects. The problems at the micro level included new business opportunities, funding, and support for the space tourism sector. The macroeconomic implications of space tourism contribute to economics and wellfare, such as the creation of jobs, the increase in tax income, the development of the supply chain, and corresponding industrial growth (Collins, 2001).

3 Methodology

While searching for and choosing publications based on titles, abstracts, and keywords, important BE themes such as risk preferences, social preferences, overconfidence, framing, restricted attention, menu effect, and social pressure concerning space tourism were selected as keywords (DellaVigna, 2009). This chapter presents a critical appraisal of the framework of BE in the field of space tourism and hospitality research by reviewing previous literature. The paper discusses the potential use of BE concepts in evaluating tourist demand for space tourism. The study draws on articles, employs inductive analysis, and leverages an understanding of the current state of BE in tourism. The findings are then heuristically applied to space tourism. The paper is a synthesis of literature searches, readings, and inductions from prior studies. Firstly, conduct a search for keywords related to "behavioural economics", such as risk preferences, overconfidence, social preferences, limited attention, framing, social pressure, as well as the menu effect. Then, search for the keywords "demand analyses and space tourism" on Google Scholar. Only include publications from the past decade (2013–2023). The literature searched is screened according to the nature of journals and the number of citations. Secondly, from these articles, it can be sorted out which concepts of the behavioral economy have been applied and how often they have been used to analyze the demand for space tourism and discuss how to apply BE concepts to study space tourism in future research in other topics.

4 Results and Discussions

In terms of space tourism, important BE concepts such as risk preferences, social preferences, overconfidence, framing, restricted attention, menu effect, and social pressure were utilized as keywords to find and pick papers based on their titles, abstracts, and keywords (DellaVigna, 2009). Based on a survey of related literature, this study suggests a review framework for BE in space tourism and hospitality research to highlight the critical elements of this domain and engage in conversation about whether BE concepts may be used to assess tourist demand in the space tourism sector. The supply chain has always been the key impediment to the commercialization of space travel. The complexity and technological focus of designing space services is at the expense of lead time and resupply present difficulties. Nonetheless, the industry is driven by public demand and the private sector's efforts to produce space equipment, plan the growth of organizations, and use a cutting-edge method of cooperation (Wooten & Tang, 2018).

This study provides an overview of market analysis and microeconomic factors in space tourism, especially for BE. Table 3 presents the result.

The above table shows related studies about space tourism and BE, or economic event or marketing. Mehran et al. (2023) argued that space tourism is worthy of in-depth research. Commercial access to space is in high demand. Still, there are serious concerns about how psychological traits and environmental factors affect individuals' reactions to space travel and whether such demands are maintainable over time. According to the idea of economic development, it can be challenging to comprehend and forecast consumer behavior as new sectors emerge (Schumpeter, 1961). According to this notion, new services must stand out in specific ways to encourage positive consumer behavior. Yet, because the space trip lacks critical components (such as competing offerings, modifications to service offerings, and market share results), customers may reject services that are theoretically and practically distinct (Al-Ansi et al., 2023). A simulation of space tourist behavior may benefit from understanding the fundamental characteristic required for arousal, which is described as "the craving for excitement and stimulation". Space tourism is a popular innovative tourism activity that most tourists are keen to partake in, which might raise their arousal and pleasantness (Wang et al., 2021).

It is highly challenging to forecast demand for goods and services in drastically new sectors since there are no reliable precedents, track records, or histories that can be utilized to anticipate future demand in space tourism (Golder & Tellis, 1997, as cited in Crouch et al., 2009). Also, rather than speculating based on people's interests or wishes, operators and regulators of space tourism would be of a crucial operational as well as financial necessity to receive a relatively precise projection of the genuine demand.

Table 3 Demand analyses and space tourism in terms of BE

Author	Topic	Main points
Crouch et al. (2009)	Modeling consumer choice behavior in space tourism	The findings from stated-preference discrete choice experiments that looked at how consumers would react to various possibilities in the developing space tourism area are presented in this study As may be predicted, there was a strong positive association between the desire to fly to space and current risk-taking behavior in recreational and leisure activities. Male and younger respondents showed much more interest in space travel
Corina Larisa Bunghez (2015)	Space Tourism Market Analysis: Current Situation and Future Trends	It has been noted that there is a sizable demand from the general population for this kind of activity. Prices will drop as more providers of these services enter the market, consumer demand will rise, and space tourism will emerge as one of the main profitable tourist industries The space tourism sector won't face competition from other tourist industries because of the product's distinctiveness, offering hitherto unrealized economic benefits
Zhang and Wang (2020)	Progress in space tourism studies: a systematic literature review	Most empirical studies focus on identifying and validating the variables affecting the choices made by space travelers. According to Laing and Crouch (2004), many demographic and behavioral factors, including cost, safety, product design, and others, will significantly influence people's views toward and interests in space tourism The type of space travel, the launch and design of the spacecraft, the location of the spaceport, the necessary training, the length of the tour, the insurance plan, and the operating company's reputation on the supply side, according to Reddy et al. (2012), can all have a significant impact on the intention of space tourism. The demand side's internal motives and risk perceptions may also have a significant effect
Danov (2020)	A Reveiw of space tourism services: Supply and demand challenges	On the one hand, space tourism offers lodging and food for a while, precisely like regular vacations. Yet, traveling into space differs from a typical tourist journey in that comfort and recreation are not factors in orbital space travel. On the other hand, no other service can match the feelings that space travel evokes. As a result, space flight is a special good with no substitute The flexibility of both the provider and the customer defines the trade-related space tourism service

(continued)

Table 3 (continued)

Author	Topic	Main points
Lei Wang et al. (2021)	The impact of tourists' perceptions of space-launch tourism: An extension of the theory of planned behavior approach	One of the most often used theories in various tourist studies to forecast a person's intentions is the theory of planned behavior model (TPB) This chapter investigates the links between attitudes, hedonic motivation, perceived behavioral control, subjective norms, and intention about space-launch tourism activities by taking the Wenchang spacecraft launch site as the tourist destination to study
Olya (2023)	Emerging space tourism business: Uncovering customer avoidance responses and behaviours	The goal to avoid has detrimental effects on both the services and tourists. This study modeled the avoidance intention of a space traveler using a "push–pull-mooring" (PPM) framework It investigated methods that describe the circumstances under which travelers avoid space tourism using configurational modeling. The pushing factor raises intentions to avoid space tourism, and the pushing as well as mooring factors lower it
Mehran et al. (2023)	Psychology of space tourism marketing, technology, and sustainable development: From a literature review to an integrative framework	This study conducted a hybrid study using a concept-driven methodology to identify the potential market segments for space tourism and to gain insights into how visitors behave Based on social-value consciousness, following sign value theory, consumer needs are viewed as cultural constructs in which consuming a certain service elevates users' social status (Cho & Kerstetter, 2004). According to Olya and Han (2023), sign value has a significant impact on people's decision to avoid space tourism since associating with those at the top of the social hierarchy may make individuals think twice about taking on the challenging task of space flight

5 Conclusions and Limitations

There are more and more studies on space tourism behavior, the space tourism market, and space tourism demand analysis from the perspectives of psychology, sociology, and behavior. The related concepts applied in current studies are risk perception, risk motive, social value consciousness, action plan theory, etc. As the space tourism market is still in its early stages and the number of tourists is limited, the demand analysis is primarily reliant on literature reviews. Additionally, some studies focus on space launch centres as their research subject. From the current research status, relevant concepts and theories of behavioral economy have begun to be applied to the analysis of space tourism market and space tourism behavior, and other concepts or research perspectives can be well applied to the analysis of space tourism phenomena in the future. To develop the space tourism market, more interdisciplinary theoretical results can be applied to the research in this field.

The type and quantity of theoretical research and empirical research on space tourism are limited, and it is unrealistic for this study to conduct a systematic literature review based on this. However, in the early stages of research development, researchers spoke freely and put forward various viewpoints, which actually made certain theoretical contributions and hints for future research.

References

Action, G. (2015). A review of behavioral economics in reproductive health.

Al-Ansi, A., Olya, H., & Han, H. (2023). Two decades of research on halal hospitality and tourism: a review and research agenda. *International Journal of Contemporary Hospitality Management, 35*(6), 2017-2055. https://doi.org/10.1108/IJCHM-12-2021-1515

Bell, D. E. (1982). Regret in decision making under uncertainty. *Operations research, 30*(5), 961-981. https://doi.org/10.1287/opre.30.5.961

Bunghez, C. L. (2015). Space tourism market analysis. Current situation and future trends. *International Conference on Marketing and Business Development, 1*(1), 97–103.

Camerer, C. (1999). Behavioral economics: Reunifying psychology and economics. *Proceedings of the National Academy of Sciences, 96*(19), 10575–10577.

Camerer, C., & Malmendier, U. (2007). Behavioral economics of organizations. *Behavioral economics and its applications, 235*, 235.

Collins, P., Iwasaki, Y., Kanayama, H., & Ohnuki, M. (1994). Commercial implications of market research on space tourism. *The Journal of Space Technology and Science, 10*(2), 2_3-2_11. https://doi.org/10.11230/jsts.10.2_3

Collins, P. (2001). Public choice economics and space policy: Realising space tourism. Acta Astronautica, 48(5-12), 921-950. https://doi.org/10.1016/S0094-5765(01)00067-4

Crouch, G. I., Devinney, T. M., Louviere, J. J., & Islam, T. (2009). Modelling consumer choice behaviour in space tourism. *Tourism Management, 30*(3), 441–454. https://doi.org/10.1016/j.tourman.2008.07.003

Danov, D. G. (2020). A review of space tourism services: Supply and demand challenges. *Journal of Tourism Leisure and Hospitality, 2*(1), 29–35.

DellaVigna, S. (2009). Psychology and economics: Evidence from the field. *Journal of Economic Literature, 47*(2), 315–372.

Fromlet, H. (2001). Behavioral finance-theory and practical application: Systematic analysis of departures from the homo oeconomicus paradigm are essential for realistic financial research and analysis. *Business economics*, 63-69.https://www.jstor.org/stable/23488166

Gabaix, X., Laibson, D., Moloche, G., & Weinberg, S. (2006). Costly information acquisition: Experimental analysis of a boundedly rational model. *American Economic Review, 96*(4), 1043-1068. https://doi.org/10.1257/aer.96.4.1043

Hanks, A. S., Just, D. R., & Brumberg, A. (2016). Marketing vegetables in elementary school cafeterias to increase uptake. *Pediatrics, 138*(2), e20151720. https://doi.org/10.1542/peds.2015-1720

Hernandez-Maskivker, G. M., Nicolau, J. L., Ryan, G., & Valverde, M. (2019). A reference-dependent approach to WTP for priority. *Tourism Management, 71*, 165-172. https://doi.org/10.1016/j.tourman.2018.10.003

Ho, T. H., Lim, N., & Camerer, C. F. (2006). Modeling the psychology of consumer and firm behavior with behavioral economics. *Journal of Marketing Research, 43*(3), 307–331.

Hursh, S. R. (1984). Behavioral economics. *Journal of the Experimental Analysis of Behavior, 42*(3), 435–452.

Koch, A., Nafziger, J., & Nielsen, H. S. (2015). Behavioral economics of education. *Journal of Economic Behavior & Organization, 115*, 3-17. https://doi.org/10.1016/j.jebo.2014.09.005

Lei Wang, C.-F.F., Wong, P. P., & Zhang. (2021). The impact of tourists' perceptions of spacelaunch tourism: An extension of the theory of planned behavior approach. *Journal of China Tourism Research*. https://doi.org/10.1080/19388160.2021.1900972

Li, S., Li, H., Song, H., & Chen, M. (2022). Mitigating tourism social costs during a pandemic: Evaluating residents' perceptions and behavioral intentions. *Journal of Travel Research, 61*(3), 493–510.

Li, S., Liu, Y., Dai, S., & Chen, M. (2022). A review of tourism and hospitality studies on behavioural economics. *Tourism Economics, 28*(3), 843–859. https://doi.org/10.1177/13548166211060889

Mahajan, J. (1992). The overconfidence effect in marketing management predictions. *Journal of Marketing Research, 29*(3), 329- 342.https://doi.org/10.1177/002224379202900304

Mehran, J., Olya, H., & Han, H. (2023). Psychology of space tourism marketing, technology, and sustainable development: From a literature review to an integrative framework. *Psychology & Marketing*. https://doi.org/10.1002/mar.21795

Neto, A. Q., Dimmock, K., Lohmann, G., & Scott, N. (2020). Destination competitiveness: How does travel experience influence choice? *Current Issues in Tourism, 23*(13), 1673–1687.

Olya, H., & Han, H. (2023). Emerging space tourism business: Uncovering customer avoidance responses and behaviours. *Journal of Vacation Marketing, 29*(3), 445-460. https://doi.org/10.1177/13567667221101

Park, S., Hahn, S., Lee, T., & Jun, M. (2018). Two factor model of consumer satisfaction: International tourism research. *Tourism Management, 67*, 82-88. https://doi.org/10.1016/j.tourman.2018.01.007

Ricciardi, V. R., & Simon, H. K. (2000). What is behavioral finance. *Business, Education and Technology Journal*.

Rind, B., & Bordia, P. (1996). Effect on restaurant tipping of male and female servers drawing a happy, smiling face on the backs of customers' checks. *Journal of Applied Social Psychology, 26*(3), 218-225. https://doi.org/10.1111/j.1559-1816.1996.tb01847

Schumpeter, J. A. (1961). Teoria do desenvolvimento econômico (pp. 81–129). Rio de Janeiro: Fundo de Cultura.

Song, H., Qiu, R. T., & Park, J. (2023). Progress in tourism demand research: Theory and empirics. *Tourism Management, 94*, 104655.https://doi.org/10.1016/j.tourman.2022.104655

Tan, H., Lv, X., Liu, X., & Gursoy, D. (2018). Evaluation nudge: Effect of evaluation mode of online customer reviews on consumers' preferences. *Tourism Management, 65*, 29–40.

Thaler, R. H., & Benartzi, S. (2004). Save more tomorrow™: Using behavioral economics to increase employee saving. *Journal of Political Economy, 112*(S1), S164–S187.

Wang, L., Stepchenkova, S., & Kirilenko, A. P. (2021). Will the present younger adults become future orbital space tourists?. *Tourism Recreation Research, 46*(1), 109-123. https://doi.org/10.1080/02508281.2020.1762342

Wilkinson, N., & Klaes, M. (2017). *An introduction to behavioral economics*. Bloomsbury Publishing.

Wooten, J. O., & Tang, C. S. (2018). Operations in space: exploring a new industry. *Decision Sciences, 49*(6), 999-1023. https://doi.org/10.1111/deci.12312

Zhang, Y., & Wang, L. (2020). Progress in space tourism studies: A systematic literature review. *Tourism Recreation Research, 47*(4), 372–383. https://doi.org/10.1080/02508281.2020.1857522

Chapter 9
How Are Macro and Micro Frameworks Integrated in Space Tourism Demand Studies?

1 Introduction

Space tourism, also known as traveling to space for leisure and recreation, has developed as a fresh component of the tourism sector, and its current development is spurring new travel options and more space exploration (Toivonen, 2022). After decades of development, space tourism is now recognized as a luxury travel experience that offers visitors thrilling private space voyages and has the potential to be marketed as a next-generation tour (Wittig et al., 2017). Space tourism may play a role in popularizing space, according to the favorable trend and advancement in this field (Forganni, 2017). The investigation of a brand-new market in the travel and tourism sector, as well as the significant positive impact on economic return, are where space tourism contributes. Not only has space emerged as a strategic domain for national defense and security, climate change, and connectivity, but the space economy is also expected to grow faster than GDP while it is also becoming increasingly important to a growing number of countries (Hurley, 2022a). Investigating the economic studies of space tourism is essential, given the economy's crucial influence on space travel.

In the process of development of space tourism, apart from industrial attention, it also drew great attention from the academic field. An increasing number of social science researchers have actively studied space tourism with different sub-domains and mainly contributed to the economic aspect over the past two decades, ranging from macroeconomics to microeconomics (Zhang & Wang, 2022). As for the research of tourism economics, it has experienced a long history, especially tourism demand studies. Based on the theories and empirical overview of progress in tourism demand research proposed by Song et al. (2023), one emerging trend lies in analyzing macro and micro frameworks integrated into exploring tourism demand. Given the handful of investigations on this topic and the vacant study of space tourism with this research idea, this study is inspired to search for potential possibilities to analyze macro and micro frameworks integrated into space tourism demand studies.Having an insight

K. Peng et al., *Space Tourism Supply Chain*, Contributions to Management Science,
https://doi.org/10.1007/978-981-95-1080-1_9

into this specific topic while combining this inspiration with space tourism and the importance of the space economy, this chapter aims to solve the question of "How are macro and micro frameworks integrated in space tourism demand studies?" Specifically, this study has three objectives: (1) to provide a sophisticated review of tourism demand and space tourism demand studies; (2) to thoroughly discuss the existing macro-economics and micro-economics theories and frameworks that had integrated in space tourism literature; (3) to explore the research gaps and future research topics for space tourism economic studies. Given that space tourism is still in early development and the lack of sufficient literature, this study employs a systematic literature review method to analyze relevant journal articles published since 2000. Specifically, descriptive analysis is firstly conducted, followed by thematic analysis. Results and discussion are provided at the end.

2 Literature Review

To thoroughly evaluate the body of literature on the study issue, reviews are undertaken by first looking into the demand for space tourism and its current state of research. This is followed by an analysis of tourist economic studies, primarily focusing on demand studies from both macro and micro perspectives. Finally, a discussion of space tourism in economic studies follows.

Space Tourism Development and Demand

Space tourism has been gone through two decades, separated from scientific space exploration in the 2000s, representing the sector of space tourism became independent. Several entrepreneurs and enterprises devoted their investment to the emergence of recreation and leisure needs for space travel. At the same time, a handful of super-rich individual tourists accomplished their orbital trips with a high expenditure. Space tourism experienced further development with technological advancement in recent years, verified by the successful launch of several suborbital spacecraft on high-profile flights (Peng et. al. (2025). These significant milestones marked that space tourism is becoming available to the public.

Reviewing the history of space tourism, even if only a small number of businesses and individuals have driven the growth of space tourism, the general public can receive the concomitant influences and appeal it provides (Spector & Higham, 2019). Although the current market is narrowed to a small segment of wealthy billionaires, its commitment pushes the further increase of space travel demand to the public (Spector & Higham, 2019). With the advancements in technology and commercialization, space tourism possesses the most significant potential to fulfill broader market demand for space tours by providing increasing opportunities for this unique experience.

Space Tourism Demand Studies

Align with the emergence and growth of space tourism, academic fields also paid increasing attention to this field of study in the past decades. The publication of space tourism-related articles, especially in tourism-oriented journals, has rapidly risen. In the process of academic research, social science scholars have actively investigated the research ideas and topics of space tourism, contributing to the formation of multi- and inner-disciplinary research themes (Zhang & Wang, 2022). Despite relative scant of space tourism literature, with the steady growth of space exploration and outer space gradually becoming a future tourism destination, many researchers in the tourism field claimed their confidence that space tourism is going to become a promising domain of study in the foreseeable future (Cater, 2010; Friel, 2020; Reddy et al., 2012).

Zhang and Wang (2022) analyzed the progress in space tourism studies from several dimensions and confirmed its increasing trend. Since the research regarding space tourism is still limited, most current studies are conducted in conceptual ways. Concerning the publications by sub-domains, the themes of space tourism can be categorized into different areas where space tourism demand is discussed. For instance, from a marketing aspect, space tourism demand is mentioned by analyzing potential travelers' attitudes, interests, intentions, motivation, and risk perceptions from the demand side (Laing & Crouch, 2004; Reddy et al., 2012). In the social-anthropology study, space tourism demand discussion can be discovered from the analysis of space tourism benefits for people and public awareness from an educational perspective (Toivonen, 2022). In political and legal issue studies of space tourism, demand is discussed in space tourism policy-making, public interests and security, and the law-making process (Abaydeldinov & Kala, 2016; Hobe, 2010). Space tourism demand studies can be mainly found in economics and finance, such as space tourism financing and funding, commercialization process, flight tour pricing, and demand estimation (Beery, 2012; Goehlich, 2005; Komerath et al., 2007). This extant research under economic topics also reveals that space tourism demand can be discussed from both micro and macro levels (Zhang & Wang, 2022), indicating the necessity and significance of investigating micro and macro frameworks integrated into space tourism demand research.

Tourism Economics and Tourism Demand

The topics of tourism economics are the allocation of resources to meet tourist demand and the effects of tourism at the macroeconomic and microeconomic levels (Sinclair et al., 2003). Both supply and demand drive tourism, so it combines these two factors to evaluate its broader effects, which include estimating the size of the industry with input–output models, calculating the impact of increased tourism demand using computable general equilibrium models, and determining how significant tourism is to economic growth (Sinclair et al., 2003; Stabler et al., 2009). The literature on the tourist supply side focuses mainly on its products and services, including vertical and horizontal integration in general areas such as lodging, food and beverage, tour agency, and transportation. Regarding tourism demand, the literature primarily investigates tourist motivation, estimation of demand responsiveness to various variables, and forecasts of demand future levels.

Assessing tourism demand has seen consistent growth from both macro and micro perspectives. The number of visitor arrivals is the main instrument for evaluating tourism demand at the macro level (Song et al., 2012). According to the study conducted by Song et al. (2023), tourist arrival rates can be determined by variations in domestic and international transportation, inbound and outbound tourism, and specific tourism activities, based on individual research objectives. With the progress in tourism research, numerous new measures have emerged, which include tourists' spending, potential occupancy of lodging facilities, hotel income, and length of stay. To predict tourism demand, GDP, exchange rate and prices are significant macroeconomic factors. Additionally, it can be considered that various other factors, including income, physical distance, economic uncertainty, international trade, social media content, political stability, and economic crisis, also play a crucial role. Additionally, as outlined in their research, individual reservation records, spending, motivation, and decisions relating to the duration and frequency of travel are all indicators for analysing tourist demand at a micro level. Demographic factors like income, age, travel companion, previous visit experiences, price, seasonality, and the evolving subject of online reviews all have a bearing on estimation. Finally, it is noteworthy that although academics have comprehensively examined tourism demand from macro and micro perspectives, there remains a research gap with regards to investigating the integration of macro and micro frameworks in tourism demand. It is therefore recommended that future research direction should focus on addressing this gap. Effective tourism demand forecasting models are essential for producing accurate estimates of tourism demand. Many different forecasting models for tourism demand have been developed in recent years. Still, the two primary types most widely used are time series analysis and machine learning models (Jiang et al., 2020). These two models were developed to examine the effects of tourism demand forecasting for various regions by learning the characteristics of time series (Zheng et al., 2022). Time series analysis, econometric methods, and models based on artificial intelligence are three other ways of demand modeling classification listed as three categories (Dong et al., 2023).

3 Methodology

By applying the systemic literature review method, this study follows the five-step analysis procedure (Pickering & Byrne, 2014), including (1) establishing research goals and objectives, (2) developing review criteria and protocols; (3) scanning and refining literature; (4) extract and analyze literature; and (5) synthesize and present related findings.

Firstly, the research aim is to analyze macroeconomics and microeconomics frameworks integrated in space tourism studies. Secondly, to develop review criteria, the term "space tourism" is used to search in databases and check through keywords, titles, and abstracts. The selection of "space tourism" is determined by the limited space tourism literature and to ensure no missing relevant articles. The academic databases selected for searching include Web of Science, Science Direct, Sage, Scopus, Emerald, ProQuest, and EBSCOhost. Given the startup of space tourism in the early 2000s, this study considers articles published between 2000 and 2023. Meanwhile, the retrieved articles are peer-reviewed journal articles in social science disciplines, specifically in tourism.

The next step of scanning and refining literature was conducted from March to April 2023. By searching the keyword "space tourism" in each academic database, the first round of potential article collection covered 852 journal articles, including 176 from Web of Science, 375 from Science Direct, 70 from Sage, 221 from Scopus, 75 from Emerald, 330 from ProQuest, and 45 from EBSCOhost. To systemically arrange and filter these articles, Endnote 20 was adopted for import and review. During the review, 265 articles were removed as duplications, remaining 587 as screened records. These remaining ones were filtered again to exclude the articles from non-social science disciplines and non-direct space tourism topics. After this process, the 452 remaining articles were ensured to have at least a component related to space tourism. After this procedure, the filtering condition was adjusted by selecting themes related to economics, finances, marketing and any other topics that may include space tourism demand studies, reaching 103 articles. Finally, 32 research were selected as the final eligible studies, which can be analyzed from either macroeconomic or microeconomic dimensions, while some are discovered to conclude both sides. The whole process of literature scanning and refining is demonstrated in Fig. 1.

To conduct data analysis, this study first applied a quantitative method by examining the descriptive data of authors, publication year, title, and journal name, as shown in Table 1. As for the qualitative method, thematic analysis was adopted to categorize these articles (Braun & Clarke, 2006). It followed the procedures from

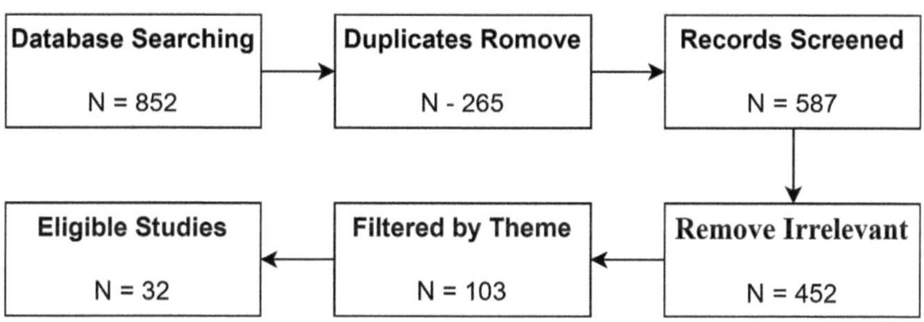

Fig. 1 Literature scanning and refining process

familiarizing the data, generating coding and themes, and confirming themes. Specifically, it started by reading the articles, especially the finding parts, and labeling them with codes of keywords concerning economic theories, frameworks, and concepts, such as tourists' motivation, expenditure, and pricing models. By transforming these codes, three themes were identified as the macro-economic domain, micro-economic domain, and both macro- and micro-domain, presented in Table 1.

4 Results and Discussions

Descriptive Analysis

As shown in Table 1, space tourism articles related to macro- and micro-economics topics range from 2000 to 2023, with steady growth over the years. The number of micro-economics articles in space tourism research is more than the ones in the macro-economics domain. Moreover, six articles overlap both topics. Given the general limitation in space tourism literature, the research regarding micro-economics and macro-economic in space tourism demand studies is still rare. As for the journals for publication, most of the selected articles are discovered in tourism or space field journals, while a few are in other social science fields. In detail, Space Policy and Acta Astronautica are the top two in space-focused journals, while tourism recreation research presents the highest frequency in space tourism economic analysis.

Thematic Analysis

Following descriptive analysis, thematic analysis investigates each domain in detail. As noted in earlier sections, economic research is a significant focus of space tourism research. Although the related literature in this area is limited, the existing economic studies in space tourism are found to be themed on economic and financial opportunities and challenges, proposing new concepts and models related to space tourism. In contrast, space tourism demand is discussed or implied under these topics. In line with these topics, illustrations of space tourism demand of each economic domain in the selected data are presented as follows.

As for the macro-level domain, space tourism demand is first indicated in articles related to discussing the economic impacts of space tourism. It includes the analysis of the expansion of other industries that are related to the space tourism industry, the influence of increasing opportunities such as job generation and taxation creation on market demand for space tourism, and the impact of supply chain adjustment on space tourism demand (Collins, 2001; Collins & Autino, 2010; Crawford, 2016; Crouch, 2001; Friel, 2020; Smith, 2000; Webber, 2013). Moreover, space tourism

Table 1 Data analysis results

Themes	Authors	Year	Title	Journal
Macro-economics	Reichert	2001	The future of human spaceflight	Acta Astronautica
Macro-economics	Collins	2002	Meeting the needs of the new millennium: Passenger space travel and world economic growth	Space Policy
Macro-economics	Collins & Autino	2010	What the growth of a space tourism industry could contribute to employment, economic growth, environmental protection, education, culture, and world peace	Acta Astronautica
Macro-economics	Valentine	2012	Exit strategy: Profit, cosmology, and the future of humans in space	Anthropological Quarterly
Macro-economics	Genta	2014	Private space exploration: A new way for starting a spacefaring society?	Acta Astronautica
Macro-economics	Crawford	2016	The long-term scientific benefits of a space economy	Space Policy
Macro-economics	Friel	2020	Tourism as a driver in the space economy: new products for intrepid travelers	Current Issues in Tourism
Macro-economics	Spector	2020	he cosmic-local nexus: Intersections between outer space and rural communities	Tourism Recreation Research
Macro-economics	Toivonen	2022	Sustainability dimensions in space tourism: the case of Finland	Journal of Sustainable Tourism
Macro-economics	Scott	2022	A space tourism destination: environmental, geopolitical and tourism branding considerations for New Zealand as a 'launch state'	Journal of Sustainable Tourism
Macro-economics	Roman, Kosiński, Kumar, Niedziółka, & Krasnodębski	2022	Virtual and Space Tourism as New Trends in Travelling at the Time of the COVID-19 Pandemic	Sustainability

(continued)

Table 1 (continued)

Themes	Authors	Year	Title	Journal
Macro-economics	Peng, Hsu, Lin, & Su	2022	Proposing spacetime scale for space tourism economics	Tourism Economics
Micro-economics	Collins and Funatsu	2000	Collaboration with aviation—The key to commercialisation of space activities	Acta Astronautica
Micro-economics	Smith	2000	Space tourism: The twenty-first century "frontier"	Tourism Recreation Research
Micro- economics	An et al	2001	Dreams and realities: The challenges facing development of space tourism	Space Policy
Micro-economics	Goehlich	2005	A ticket pricing strategy for an oligopolistic space tourism market	Space Policy
Micro-economics	Polkowska	2021	Space Tourism Challenges	Review of European and Comparative Law
Micro-economics	Dewily & Michael	2021	Space Tourism Activities Overview of International Law	Journal of International Trade, Logistics and Law
Micro-economics	Bogdal-brzezi & Wendt	2021	Space Tourism—Between Competition and Cooperation of States and Non-State Entities	GeoJournal of Tourism & Geosites
Micro-economics	Zhang & Wang	2022	Progress in space tourism studies: a systematic literature review	Tourism Recreation Research
Micro-economics	Martin & Freeland	2022	A Round Trip to the Stars? Considerations for the Regulation of Space Tourism	Air & Space Law
Micro-economics	Khan, M. A	2022	An Analysis of the Space Tourism Market in the United Arab Emirates and the Kingdom of Saudi Arabia and Its Potential for Development of Zero-Gravity and Suborbital Commercial Spaceflights	New Space-the Journal of Space Entrepreneurship and Innovation

(continued)

Table 1 (continued)

Themes	Authors	Year	Title	Journal
Micro-economics	Mesa-Arango, Pineda-Jaramillo, Araujo, Bi, Basva, & Viti	2023	Missions and factors determining the demand for affordable mass space tourism in the United States: A machine learning approach	Acta Astronautica
Micro-economics	Rees, Ryden, Hands, & Clewer	2023	Radiation risk assessment for varying space weather conditions for very high altitude 'near space' tourism balloon flights	Journal of Space Safety Engineering
Micro-economics	Lemarqu, Welch, & Iliev	2023	Business Case Study for the Cycler: A Circumlunar Vehicle for Development of Space Tourism and Lunar Infrastructure by 2030	New Space-the Journal of Space Entrepreneurship and Innovation
Micro-economics	Peng, Chan, & Wu	2023	Space tourism flow generated from social media data	Data in Brief
Macro- and micro-economics	Collins	2001	Public choice economics and space policy: Realizing space tourism	Acta Astronautica
Macro- and micro-economics	Crouch	2001	The market for space tourism: Early indications	Journal of Travel Research
Macro- and micro-economics	Komerath et al	2007	Policy model for space economy infrastructure	Acta Astronautica
Macro- and micro-economics	Ingle	2011	The Copernican shift in space tourism and its implications for tourism in the Great Karoo	South African Geographical Journal
Macro- and micro-economics	Beery	2012	State, capital, and spaceships: A terrestrial geography of space tourism	Geoforum
Macro- and micro-economics	Webber	2013	Space tourism: Its history, future, and importance	Acta Astronautica

market demand is estimated to grow based on the terrestrial destinations' development in the neighborhood area of spacecraft launching places (Friel, 2020; Spector, 2020). Space tourism destination promotion is also discussed in tourism branding, suggesting the relevant transformation of inbound tourists' market segment and needs (Scott, 2022). Furthermore, the framework of neoliberal capitalism is raised as a new research direction, implying the potential demand for space tourism from newly generated tourism markets (Valentine, 2012). Space tourism is also investigated from the sustainability development aspect, which mentions its impact on the change of demand structure (Toivonen, 2022). Additionally, recent research developed a space-time scale that derived an economic model for space tourism based on the General Relative Theory (Peng et al., 2022). Finally, space tourism demand is also discussed in the topic of Covid-19, since the emergence of space travel generates a new market trend (Roman et al., 2022).

In terms of the micro-economics domain, the body of extant research demonstrates an emphasis on new avenues for businesspeople to finance and fund space tourism privately, in contrast to past partnerships or collaborations with the government (Beery, 2012; Bógdał-Brzezińska & Wendt, 2021; Collins, 2002; Collins & Funatsu, 2000). Economic studies on space tourism include assessing the cost of space flight tickets and the revenue stream (Goehlich, 2005). Besides, space tourism demand on the micro-level is estimated from a social-economic perspective by proposing organized and responsible regulations for protecting and managing individuals' needs in space travel (Dewily & Michael, 2021; Martin & Freeland, 2022). Furthermore, scholars conducted research on analyzing the actual space tourism market demand in specific regions, achieved by quantitative method of collecting data such as attitudes, behaviors, and willingness to pay (Khan, 2022) and travelers' choices with a machine learning approach (Mesa-Arango et al., 2023). Another aspect of examining space tourism demand is assessing space travel's potential risk and travelers' perception of risk (Rees et al., 2023). Meanwhile, researchers also proposed a traveler-focused cislunar transportation system as the chance for investors and travelers' participation in space tourism, which can also be applied in defining market demand (Lemarquis et al., 2023). Moreover, using online data collection from potential space travelers, a quantitative approach has proven helpful in space tourism studies, indicating its effectiveness in analyzing space tourism demand in future studies (Peng, Chan et al., 2023).

For the data containing both macro- and micro-economics analysis, Collins (2001) discussed space tourism demand from the macro-level of the public choice economy and the micro-level of new opportunities for the private market sector. Crouch (2001) analyzed the forecast of space tourism demand by summarizing all related research from both sides of interest bodies, including public organizations and individual commitment. It is also suggested by Komerath et al. (2007) that space tourism can indirectly benefit from broadening space travel commercialization and facilitating space infrastructure construction from a macroeconomic perspective while proposing a profit stream framework for estimating the business performance of future space hotels. Another research regarding space tourism demand is delineated by analyzing the spacecraft launch site construction from a micro angle and its positive impact on terrestrial tourism destinations from a macro angle (Ingle, 2011). Moreover, Beery (2012) claimed that space tourism demand is associated with entrepreneurship opportunities that space tourism brings to public and political-economic geography as a brand-new space tourism research idea. Lastly, Webber (2013) explored space tourism demand by examining its economic benefit on scale effects from the micro-level and the enlargement of other industries related to the space tourism sector from the macro-level.

Discussions

The literature, spanning 2000 to 2023, shows a steady growth in articles, with more micro-economic studies than macro-economic, and some overlap in both domains. Key journals include Space Policy, Acta Astronautica, and Tourism Recreation Research.

Macro-Economics

Demand is discussed in terms of economic impacts, such as job generation and taxation (Collins & Autino, 2010; Crawford, 2016), and the influence on related industries. Destination promotion and sustainability are also key, with studies like Scott (2022) suggesting transformations in market segments. Recent research, such as Peng, Hsu et al. (2023), proposes a spacetime scale for economic modeling, and Roman et al. (2022) discuss COVID-19 impacts. These studies paint a multifaceted picture of space tourism demand in macroeconomics. It's a lever for growth (jobs, taxation, industry spillovers), a target for strategic promotion (space tourism types, sustainability), a subject of innovative modeling (spacetime scales), and a variable reshaped by global crises (COVID-19). The research collectively suggests that understanding and managing space tourism demand research requires both granular analysis—specific industries, regions, or timeframes—and a holistic view of its broader economic echoes. Policymakers, businesses, and economists might use these insights to navigate challenges like climate goals, or technological disruption, ensuring space tourism demand drives prosperity without destabilizing the system.

Micro-Economics

Emphasis is on private funding, ticket pricing, and regulations (Dewily & Michael, 2021; Goehlich & Rücker, 2005). Market demand is analyzed in specific regions using quantitative methods, such as willingness to pay and machine learning (Khan, 2022; Mesa-Arango et al., 2023). Risk assessment and traveler-focused systems are also explored (Idrees et al., 2023; Lemarquis et al., 2023). These studies painted a picture of a micro-economic framework where private firms navigate funding, pricing, and regulatory challenges to meet regional demand. Quantitative tools like willingness-to-pay surveys and machine learning provide data-driven insights into consumer behavior, while risk assessment ensures sustainability. The cited studies span nearly two decades (2005–2023), suggesting a blend of foundational theories and cutting-edge techniques, likely applied to a dynamic, consumer-facing industry for sapce tourism demand research.

Integrated Macro and Micro Economics Studies

Studies like Collins (2001)and Crouch (2001) integrate public and private sector perspectives, forecasting demand and discussing economic benefits from scale effects and industry enlargement (Komerath et al., 2007; Webber, 2013). In summary, the research content underscores the importance of integrating macro- and micro-economics perspectives to forecast space tourism demand and understand the economic benefits of scale effects and industry enlargement. These studies provide valuable insights for policymakers, industry leaders, and researchers aiming to foster sustainable economic growth and development of space tourism.

5 Conclusions

Aimed at investigating how macro and micro frameworks are integrated in space tourism demand studies, this article adopted a systemic literature review approach, combining qualitative and quantitative methods in analyzing 32 articles related to the research theme. Through scanning and reviewing the database coding the keywords into each category, the eligible articles are demonstrated in detail under the domains of macro-economics, micro-economics, and macro- and micro-economics in space tourism.

The results show that in macro-economic studies related to space tourism, demand is predicted through space tourism's social-economic benefits on supply chain and related industrial enlargement, taxation generation, job creation, space commercialization and infrastructure construction process, terrestrial tourism destinations marketing, political-economic geography, sustainability development through space tourism and economic benefits during Covid-19. As for the microeconomic field, estimated space travel demand by assessing motivation, opportunities for individual sectors, financing and funding for the space tourism industry, public–private collaboration, ticketing pricing, profit stream, economy of scale effects, and quantitative data collection methods. Articles containing both macro and micro economic sides overlap the topics with other studies regarding either macro or micro, indicating a promising future research direction.

In terms of the theoretical contribution of this study, it is the first research to analyze economic frameworks in space tourism demand from both macro and micro perspectives, filling the research gap in this vacant field of study. Meanwhile, it can also provide theoretical support for future research. On the industrial side, it helps industry practitioners better understand the demand for space tourism to customize their services to transportation agencies that require guidance to support fund management related to commercial space exploration.

Regarding suggestions for future research in space tourism, since most of the current research of space tourism demand studies are conceptual, it is suggested that empirical studies should also be well-noticed and conducted. Another future research idea concerns research on tourism demand in different market segments.

Since existing ones mainly focus on the wealthy market or make assumptions about the general public market, future research could categorize the potential markets into different segments to better understand consumers and their needs.

Referring to the limitation, the main issue of this chapter is regarding data selection. The data adopted in this chapter only considered peer-reviewed journal articles, while only the literature in the social science field is included. However, research in other disciplines, such as science and technology, may also provide valuable viewpoints and support. Besides, similar types of tourism, such as VR tourism and additional space-related experiential tourism, are also helpful for space tourism demand study.

References

Abaydeldinov, Y., & Kala, N. (2016). International legal aspects of tourism activity: International treaties analysis. *Journal of Advanced Research in Law and Economics, 7*(4(18)), 714–720.

Beery, J. (2012). State, capital and spaceships: A terrestrial geography of space tourism. *Geoforum, 43*(1), 25–34. https://doi.org/10.1016/j.geoforum.2011.07.013

Bógdał-Brzezińska, A., & Wendt, J. A. (2021). Space tourism-between competition and cooperation of states and non-state entities. *Geo Journal of Tourism and Geosites, 38*(4), 1151–1156. https://doi.org/10.30892/gtg.38421-755

Braun, V., & Clarke, V. (2006). Using thematic analysis in psychology. *Qualitative Research in Psychology, 3*(2), 77–101. https://doi.org/10.1191/1478088706QP063OA

Cater, C. I. (2010). Steps to space; Opportunities for astrotourism. *Tourism Management, 31*(6), 838–845. https://doi.org/10.1016/j.tourman.2009.09.001

Collins, P. (2001). Public choice economics and space policy: Realising space tourism. *Acta Astronautica, 48*(5–12), 921–950.

Collins, P. (2002). Meeting the needs of the new millennium: Passenger space travel and world economic growth. *Space Policy, 18*(3), 183–197. https://doi.org/10.1016/S0265-9646(02)00018-8

Collins, P., & Autino, A. (2010). What the growth of a space tourism industry could contribute to employment, economic growth, environmental protection, education, culture and world peace. *Acta Astronautica, 66*(11–12), 1553–1562. https://doi.org/10.1016/j.actaastro.2009.09.012

Collins, P., & Funatsu, Y. (2000). Collaboration with aviation—The key to commercialisation of space activities. *Acta Astronautica, 47*(2–9), 635–646.

Crawford, I. A. (2016). The long-term scientific benefits of a space economy. *Space Policy, 37*, 58–61.

Crouch, G. I. (2001). The market for space tourism: Early indications. *Journal of Travel Research, 40*(2), 213–219. https://doi.org/10.1177/004728750104000212

Dewily, R. D. A., & Michael, T. (2021). Space tourism activities overview of international law. *Journal of International Trade, Logistics and Law, 7*(1), 8–12.

Dong, Y., Xiao, L., Wang, J., & Wang, J. (2023). A time series attention mechanism based model for tourism demand forecasting. *Information Sciences, 628*, 269–290. https://doi.org/10.1016/j.ins.2023.01.095

Forganni, A. (2017). The potential of space tourism for space popularisation: An opportunity for the EU Space Policy? *Space Policy, 41*, 48–52. https://doi.org/10.1016/j.spacepol.2017.04.005

Friel, M. (2020). Tourism as a driver in the space economy: New products for intrepid travellers. *Current Issues in Tourism, 23*(13), 1581–1586. https://doi.org/10.1080/13683500.2019.1628189

Goehlich, R. A. (2005). A ticket pricing strategy for an oligopolistic space tourism market. *Space Policy, 21*(4), 293-306. https://doi.org/10.1016/j.spacepol.2005.08.007

Goehlich, R. A., & Rücker, U. (2005). Low-cost management aspects for developing, producing and operating future space transportation systems. *Acta Astronautica, 56*(1–2), 337–346. https://doi.org/10.1016/j.actaastro.2004.09.010

Hobe, S. (2010). The legal regime for private space tourism activities—An overview. *Acta Astronautica, 66*(11–12), 1593–1596.

Idrees, H., Xu, J., Haider, S. A., & Tehseen, S. (2023). A systematic review of knowledge management and new product development projects: Trends, issues, and challenges. *Journal of Innovation & Knowledge, 8*(2), Article 100350. https://doi.org/10.1016/j.jik.2023.100350

Jiang, P., Yang, H., Li, R., & Li, C. (2020). Inbound tourism demand forecasting framework based on fuzzy time series and advanced optimization algorithm. *Applied Soft Computing, 92*, Article 106320.

Khan, M. A. (2022). An analysis of the space tourism market in the United Arab Emirates and the Kingdom of Saudi Arabia and its potential for development of zero-gravity and suborbital commercial spaceflights. *New Space, 10*(4), 315–329.

Komerath, N., Nally, J., & Tang, E. Z. (2007). Policy model for space economy infrastructure. *Acta Astronautica, 61*(11–12), 1066–1075. https://doi.org/10.1016/j.actaastro.2006.12.033

Laing, J. H., & Crouch, G. I. (2004). Australian public interest in space tourism and a cross-cultural comparison. *Journal of Tourism Studies, 15*(2), 26–36.

Lemarquis, N., Welch, C., & Iliev, Y. (2023). Business case study for the cycler: A circumlunar vehicle for development of space tourism and lunar infrastructure by 2030. *New Space, 11*(1), 58–67.

Martin, A.-S., & Freeland, S. (2022). A round trip to the stars? Considerations for the regulation of space tourism. *Air and Space Law, 47*(2). https://doi.org/10.54648/aila2022014

Mesa-Arango, R., Pineda-Jaramillo, J., Araujo, D. S. A., Bi, J., Basva, M., & Viti, F. (2023). Missions and factors determining the demand for affordable mass space tourism in the United States: A machine learning approach. *Acta Astronautica, 204*, 307–320. https://doi.org/10.1016/j.actaastro.2023.01.006

Peng, K.-L., Chan, X., & Wu, C.-H. (2023). Space tourism flow generated from social media data. *Data in Brief, 48*, Article 109061.

Peng, K.-L., Hsu, C. H., Lin, P. M., & Su, M. (2022). Proposing spacetime scale for space tourism economics. *Tourism Economics*, 13548166221109666.

Peng, K.-L., Hsu, C. H. C., Lin, P. M. C., & Su, M. (2023). Proposing spacetime scale for space tourism economics. *Tourism Economics, 29*(6), 1671–1678. https://doi.org/10.1177/13548166221109666

Peng, K. L., Kim, M. J., Huang, Z., Lai, I. K., & Ji, X. (2025). From holocene to anthropocene: A spacetime perspective on the ontology, epistemology, and methodology of space tourism. *Journal of Hospitality and Tourism Management,* 101304. https://doi.org/10.1016/j.jhtm.2025.101304

Pickering, C., & Byrne, J. (2014). The benefits of publishing systematic quantitative literature reviews for PhD candidates and other early-career researchers. *Higher Education Research & Development, 33*(3), 534–548. https://doi.org/10.1080/07294360.2013.841651

Reddy, M. V., Nica, M., & Wilkes, K. (2012). Space tourism: Research recommendations for the future of the industry and perspectives of potential participants. *Tourism Management, 33*(5), 1093–1102. https://doi.org/10.1016/j.tourman.2011.11.026

Rees, C., Ryden, K., Hands, A., & Clewer, B. (2023). Radiation risk assessment for varying space weather conditions for very high altitude 'near space' tourism balloon flights. *Journal of Space Safety Engineering.*

Roman, M., Kosiński, R., Bhatta, K., Niedziółka, A., & Krasnodębski, A. (2022). Virtual and space tourism as new trends in travelling at the time of the COVID-19 pandemic. *Sustainability, 14*(2), 628.

Sinclair, M. T., Blake, A. D. A. M., & Sugiyarto, G. U. N. T. U. R. (2003). The economics of tourism. *Classic reviews in tourism*, 22-54.
Stabler, M. J., Papatheodorou, A., & Sinclair, M. T. (2009). *The economics of tourism*. Routledge.https://doi.org/10.4324/9780203864272
Scott, M. (2022). A space tourism destination: Environmental, geopolitical and tourism branding considerations for New Zealand as a 'launch state.' *Journal of Sustainable Tourism, 30*(9), 2240–2253. https://doi.org/10.1080/09669582.2020.1817049
Smith, V. L. (2000). Space tourism: The 21st century "frontier." *Tourism Recreation Research, 25*(3), 5–15.
Song, H., Dwyer, L., Li, G., & Cao, Z. (2012). Tourism economics research: A review and assessment. *Annals of Tourism Research, 39*(3), 1653–1682. https://doi.org/10.1016/j.annals.2012.05.023
Song, H., Qiu, R. T., & Park, J. (2023). Progress in tourism demand research: Theory and empirics. *Tourism Management, 94*, 104655. https://doi.org/10.1016/j.tourman.2022.104655
Spector, S. (2020). Delineating acceptable risk in the space tourism industry. *Tourism Recreation Research, 45*(4), 500–510.
Spector, S., & Higham, J. E. S. (2019). Space tourism in the Anthropocene. *Annals of Tourism Research, 79*, Article 102772. https://doi.org/10.1016/j.annals.2019.102772
Toivonen, A. (2022). Sustainability dimensions in space tourism: The case of Finland. *Journal of Sustainable Tourism, 30*(9), 2223–2239. https://doi.org/10.1080/09669582.2020.1783276
Valentine, D. (2012). Exit strategy: Profit, cosmology, and the future of humans in space. *Anthropological Quarterly*, 1045–1067.
Webber, D. (2013). Space tourism: Its history, future and importance. *Acta Astronautica, 92*(2), 138–143.
Wittig, M. C., Sommerrock, F., Albers, M., & Beil, P. (2017). *Rethinking luxury: How to market exclusive products and services in an ever-changing environment*. LID Editorial.
Zhang, Y. Z., & Wang, L. R. (2022). Progress in space tourism studies: A systematic literature review. *Tourism Recreation Research, 47*(4), 372–383. https://doi.org/10.1080/02508281.2020.1857522
Zheng, H., Fu, L., & Ye, Q. (2022). Flexible capped principal component analysis with applications in image recognition. *Information Sciences, 614*, 289–310.

Chapter 10
How Did COVID-19 Affect Space Tourism Demand?

1 Introduction

COVID-19 created unprecedented public fear, and tourism was one of the areas that suffered the most during the COVID-19 pandemic outbreak. When the pandemic began, various countries imposed travel restrictions. For example, some provinces in China suspended public transportation, closed schools and assemblies, and closed border crossings altogether. According to the United Nations World Tourism Organization report published in 2020, there is an estimated risk of 100–120 million tourism jobs due to the pandemic. The report suggests a decline of over 80 percent in global tourism, and the global economy stands to lose up to $1.2 billion. Meanwhile, scholars in the field of tourism have been quick to respond to COVID-19 and have published numerous research-based articles in this context. Research shows that COVID-19 is already having a severe impact on tourism stakeholders. In the case of individual tourists, COVID-19 has increased the perceived risk and threat of tourism and created fear of tourism (Zhang & Wang, 2022).

When individuals are behaviorally restricted, it inspires people to seek creative and travel forms. Because when individuals feel tired, stressed, and anxious, they need something to distract them and refresh their minds. This encourages them to find alternatives to real-world interests and hobbies under the influence of COVID-19. In this context, space tourism grew rapidly during COVID-19, especially virtual space tourism. Many people, unable to function under the influence of COVID-19, shifted their activities to the "virtual world." For example, concerns about the epidemic during COVID-19 accelerated the boom in virtual communities, including online gaming and virtual reality tourism. Some experts predict that the space tourism market will grow at an average annual rate of 16.9%, with revenues for related companies reaching $1.3 billion by 2025 (Roman et al., 2022). Among these, virtual space tourism, a cross between space tourism and virtual tourism, has also attracted the attention of researchers. Studies have been conducted to show the evolution of virtual space tourism applications (Roman et al., 2022). However, most studies have been

K. Peng et al., *Space Tourism Supply Chain*, Contributions to Management Science,
https://doi.org/10.1007/978-981-95-1080-1_10

based on conceptualization, and few have focused on the mechanisms of external environmental stimuli influencing space tourism. Despite the growing interest of researchers in COVID-19, it seems to be explored around traditional tourism destinations. Few studies have focused on the impact on space tourism demand under the influence of COVID-19. Virtual space tourism, as a type of space tourism, is also only tip of the iceberg. Therefore, this chapter proposes to use the space tourism research literature published during COVID-19 in the Web of Science database as a sample data source to map the knowledge in the field of space tourism research with the help of CiteSpace software, an information visualization and analysis tool. In the theoretical sense, this study shows the research power and hotspots of space tourism research during COVID-19, explains the impact of COVID-19 on space tourism demand analysis based on the literature research perspective, and provides a comprehensive grasp of the research progress of international scholars on space tourism under the influence of the epidemic and provides practical and adequate references. In a practical sense, this study proposes suggestions for space tourism development in the post-epidemic period.

2 Literature Review

Challenge to Tourism Demand in COVID-19

The link between demand crises and tourism has gained momentum in recent years, increasing public awareness of the threat of global terrorism or economic crisis. In times of crisis, the value and flow of tourism between the origin and destination regions can be directly affected. In addition, crises can lead to significant changes in demand for specific destinations. Given the tourist's background, perceived risk can influence tourist behavior and destination choice (Reichel et al., 2007).

The COVID-19 pandemic significantly changed consumer psychology and behavior (Huang & Wang, 2023). The epidemic adversely affected tourists' behavior and mental health, including anxiety, fear, and sadness (Yin et al., 2022). Potential tourists may also abandon planned travel itineraries due to fear of disease infection, as there appears to be no way to avoid virus transmission during travel (Abbas et al., 2021). At the same time, mobility against COVID-19 has had a significant impact due to direct restrictions or bans on transportation services, closure of activities, and imposition of self-imposed limits. Many studies have reported that users' habits have changed significantly worldwide, with different travel frequencies and destinations (Politis et al., 2021). In Germany, about 60% of people have changed their mode of transportation, especially the number of unimodal trips has increased from 68 to 83%; in the Netherlands, about 80% of people have reduced their outdoor activities and traveled 55% less compared to 2019. It should be noted that the tremendous development of transportation means has facilitated interactions between cultures, with business and tourism being among the most well-known purposes of these interactions. During a new viral outbreak, one can expect a significant reduction in travel, which will ultimately affect the supply and demand for travel-related products and services.

Space Tourism and Virtual Space Tourism

Space tourism, strictly speaking, is "travel and temporary residence in space beyond the Earth's atmosphere". The "Carmen Line" set by the International Aeronautical Federation (IAF) is considered the boundary between the Earth's atmosphere and space, at an altitude of 100 km, above which a vehicle can fly around the Earth for more than a week without much consideration for aerodynamics and using gravity as a centripetal force. Suppose the flight altitude is lower than the Carmen line. In that case, the air resistance will be lost, depleting the vehicle's kinetic energy and reducing flight speed. The car cannot maintain the flight altitude for a long time. In summary, space tourism is a particular type of space activity, with sightseeing and excursions as the primary purpose, and the minimum altitude should be above 100 km.

Mehran et al., (2023) points out that space tourism is one of the categories of adventure and special activities, arguing that the first adventure in space tourism occurred in 2001 when the first commercial tourist, Dennis A. Tito, spent $20 million on April 28, 2001, to experience space tourism and stayed on the International Space Station for seven days. Before that, only government astronauts were allowed to go into space. Seven tourists have traveled to the ISS on Russian Soyuz rockets and spacecraft, making eight trips to near-Earth orbit. In 2004, Laing and Crouch proposed five categories for the space tourism market: (1) virtual—virtual reality headsets that allow people to get a bird's eye view of space; (2) terrestrial—combining space camps, theme parks, space museums, and other activities to provide a space experience without leaving the surface of the Earth; (3) near space—very high supersonic cruises that allow people to experience short periods of weightlessness and high altitude tours, etc. (4) suborbital—taking people high enough to view the curvature of the Earth and the blackness of space without leaving Earth's orbit; and (5) LEO/high Earth orbit—actual Earth orbit experiences that last from a few hours to a few days or even weeks. Considering physical distance alone, the Earth is relatively close in space. The competition between three billionaires (Elon Musk, Jeff Bezos, and Richard Branson) has likewise accelerated the commercialization of space tourism. This study focuses on virtual space tourism among the types of space tourism.

Despite the high cost of space tourism, it has been considered a form of tourism for which the rich and adventurous are willing to pay a high price. But space tourism has become a fact of life, and as Reddy et al., (2012) describes, a rapid growth phase occurs when the cost of space tourism decreases due to mature technology, economies of scale, and competition. Finally, a saturation phase (i.e., civilian mass space tourism) occurs when economic expansion and new markets drive growth.

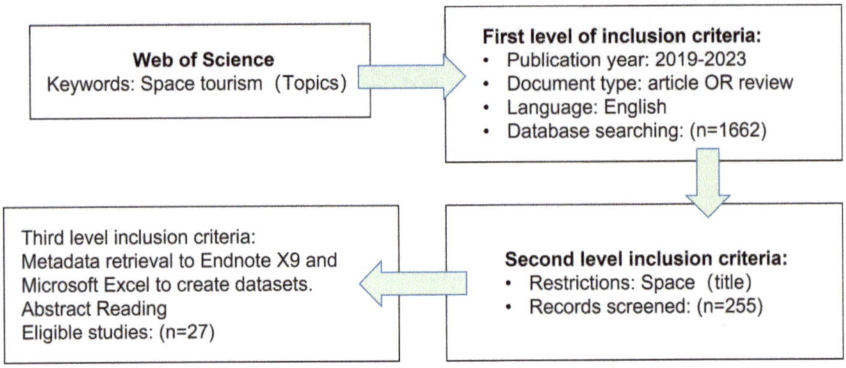

Fig. 1 Literature data collection process

However, in its current form, it will still be some time before space tourism goes civilian. It is reassuring to note that the "International Space Station VR Tour" was released in 2018, an app allowing users to view the interior of the International Space Station. Then, Forward Thinking Interactive's "Spacewalk VR Experience" app, released in March 2021, allows owners of VR glasses to walk through space without leaving home and will enable people to select a day or night part of the scenario. One can also choose a part of the scene to perform some maintenance on the space shuttle (as shown in Fig. 1). Virtual space tourism is becoming increasingly popular as a digital platform for interaction, socialization, and exploration in virtual environments, offering an alternative to real-time travel.

3 Methodology

This study employed Pickering and Byrne's (2014) literature review process to provide a comprehensive analysis of research on space tourism during COVID-19, with Fig. 1 displaying the review process.

After collecting the data, the author utilized CiteSpace to visualize and analyze the literature. This software is an information visualization tool primarily designed for measuring and analyzing data from scientific literature. It can assess literature specific to a certain domain to uncover the key pathways of disciplinary domain evolution and its inflection points of knowledge. Additionally, it can conduct an analysis of potential dynamic mechanisms of disciplinary evolution, and enable the detection of disciplinary development frontiers through a series of visualization maps.Using CiteSpace software, this study mapped the time zone distribution of keyword co-occurrence, author collaboration and research institution network mapping, and visual literature management to facilitate a more intuitive understanding of space tourism trends and development direction during COVID-19. A detailed explanation is provided by the author in the next section.

Table 1 Year and number of articles published

Year	Number of publications
2020	7
2021	6
2022	10
2023	4

4 Analysis Results and Discussions

Timing Analysis

The annual number of publications on space tourism research during COVID-19 is shown in Table 1. In terms of the amount of literature on related topics, overall, the amount of literature on space tourism research has fluctuated and increased since the outbreak of COVID-19.

Researcher Analysis

Analysis of Issuing Organizations

Figure 2 shows the prominent publishing institutions in space tourism research during COVID-19. From the knowledge map, it can be seen that the primary publishers of space tourism research are universities, with the City University of Macau, the University of Canterbury, and the University of Sheffield being the core of space tourism research output under the impact of the epidemic, and leading the research in this field. However, collaboration among research institutions across regions is relatively insufficient, and inter-institutional cooperation often exists within the same university or region, which has led to relatively slow progress in space tourism research.

Author Co-Occurrence Knowledge Graph Analysis

Figure 3 shows the co-occurrence map of authors of space tourism research publications during the epidemic, with each node representing one publication author. The larger the node (author's name), the greater the number of articles by that author; the thicker the line between authors, the stronger the collaboration or connection between authors. The analysis shows that the whole network has a core–edge structure. Peng Kang-Lin, Hossein Olya, and Heesup Han have relatively more articles, with 2, 3, and 3, respectively, and are the three most prominent nodes (author names) in Fig. 3. Beyond the core part is a peripheral circle formed by many research groups

Fig. 2 Collaborative mapping of space tourism research institutions during COVID-19

or independent authors, mostly in isolation, forming only three author groups, among which the author group led by Hossein Olya and Heesup Han has a more significant number of publications.

Fig. 3 Co-present mapping of authors of space tourism research institutions during COVID-19

The principal research authors' studies are as follows: Peng Kang-Lin specializes in space tourism economics, with research areas including the space tourism economics model, space tourism business chain, and analysis of space tourism behavioral intentions (Peng et al., 2023). The novelty of his research themes, extensive scope, and highly innovative content make him a valuable contributor to the field. His attention to the latest developments in space tourism research provides valuable insights for future research.

Hossein Olya and Heesup Han both authors focused on space tourism marketing during COVID-19, including the perceived risks of space tourism and the antecedents of space tourism behavioral intentions (Olya & Han, 2023), the psychology of space tourism sustainability (Mehran et al., 2023). These two authors collaborated during the epidemic with outstanding results, using a quantitative research approach to study space tourists based on a psychological perspective (including emotions and behaviors), and subsequently contributing to space tourism.

Research Hotspot Analysis

From Fig. 4 and Table 2, it is evident that space tourism was the most frequently researched topic during the COVID-19 pandemic with a frequency of 14. keywords such as space tourism future, challenges, policies, and virtual reality also become more significant nodes in the graph due to their more frequent occurrence. Combined with other keywords with frequencies greater than or equal to 2, it can be found that the relevant content can cover the main aspects of space tourism research during COVID-19. According to the results of CiteSpace keyword statistics and the interpretation of related literature, the research hotspots in this field can be summarized and explained from two aspects: space tourism resources and development and virtual space tourism development (see Table 3).

The main elements of the research on virtual space tourism during COVID-19 are as follows:

The first is Roman et al.'s (2022) study published in the journal Sustainability examines the emergence of virtual and space tourism as novel forms of travel during the COVID-19 pandemic. According to the authors, the combination of virtual and space tourism proved to be a creative and innovative way to explore destinations during the pandemic. The findings of the study imply that individuals utilize VR and augmented reality to suggest the NASA space station as a viable destination for tourism. VR offers a substitute form of tourism that is particularly useful during periods of crisis and pandemics, such as the COVID-19 pandemic. A majority of the respondents, more than 26%, expressed satisfaction with the perceived needs of present-day tourists in VR tours. Additionally, over 87% of the respondents share the belief that VR tourism is not a long-term substitute for real-world tourism.

Second, Zaman et al. (2022) published an article in the journal Sustainability. This studyis a valuable demonstration of empirical research in the field of virtual space travel readiness. The SEM analysis of 386 questionnaires revealed that COVID-19

Fig. 4 Keyword co-occurrence mapping for space tourism research during COVID-19

Table 2 Keywords with 2 or more occurrences of space tourism research during COVID-19

Keyword	Frequency	Keyword	Frequency
Space tourism	14	Opportunity	2
Future	5	Prospect	2
Challenge	2	Perspective	2
Determinant	2	Motivation	2
Policy	2	Virtual reality	2
Tourism	2		

Table 3 Research hotspots for space tourism during COVID-19

Research hotspot	Keyword
Space Tourism Resources and Development	Space tourism, Future, Challenge, Policy, Prospect, Opportunity
Virtual Space Tourism Development	Virtual reality, Tourism, Motivation, Determinant, Perspective

travel anxiety has a significant positive effect on preparation for virtual space travel. Furthermore, fear of missing out on travel positively moderates the relationship between COVID-19 travel anxiety and preparation for virtual space travel. Finally, tourist's technological proficiency fully mediates the relationship between COVID-19 travel anxiety and preparation for virtual space travel.

Finally, Holt (2022) published a paper in Acta Astronautica entitled "Virtual reality, augmented reality, and mixed reality: For astronaut mental health; and space tourism, education and outreach." The article presents a qualitative literature review investigating the possible application of VR/AR/MR technologies to improve the mental health of astronauts in space.

5 Conclusions

This chapter uses relevant literature in Web of Science as a data source, uses the literature review method, and uses the scientometric software CiteSpace to draw a knowledge map of space tourism research, and systematically analyzes the chronology of publications, research power, and research hotspots in this research area. The results show that (1) the number of literature on space tourism research tends to fluctuate and increase with the year, (2) Peng Kang-Lin, Hossein Olya, and Heesup Han are the scholars with the highest number of publications within the statistical range, and although several clusters of scholars are formed, there is less collaboration and communication among these scholars in general. Similarly, the universities, which are the prominent issuing institutions, do not have close academic ties with each other. (3) From the viewpoint of research hotspots, the research on space tourism in COVID-19 mainly focuses on space tourism resources and development and virtual space tourism development, primarily based on conceptual thinking of space tourism, lacking convincing empirical research.

Based on the literature on space tourism in the context of COVID-19, this chapter proposes a model of the impact on tourism demand under the stimulus of COVID-19. In this model, we consider COVID-19 as an influencing factor, which consists of internal and external factors. Internal factors refer to the risk of transmission of COVID-19 itself or isolation due to close contact with others. In contrast, external factors include government policies, epidemic trends, tourism supply, and ongoing media coverage unrelated to the COVID-19 virus. These two aspects constitute the influencing factors of COVID-19 and thus impact the tourist demand factor. At the level of tourist demand, we can explain it at the psychological and behavioral levels. At the psychological level, tourists feel psychological problems such as depression, fear, and dread and need a series of means to release negative emotions. The behavioral dimension can be understood in terms of behavioral performance, such as that tourists cannot leave their homes, socialize offline, and are limited in their previous behavioral tendencies. However, staying home will stimulate the potential tourist's need for other types of tourism at home, such as virtual space tourism, a form of tourism that is not limited by time and space and allows the tourist to experience virtual space tourism in an immersive way.

It must be noted that virtual tourism behaviors stimulate tourists' behavioral intentions in the actual environment; therefore, this study proposes a pathway for the final relationship between virtual space tourism and actual space tourism, and it is reasonable to infer that short-term adaptive tourism behaviors may evolve into long-term tourism behaviors, such as a preference for virtual space tourism.

This chapter offers personal insights on COVID-19 to accelerate the development of space tourism. First, global epidemics have led to significant changes in tourism demand. Tourists are susceptible to perceived risks in terms of health and economics, and the epidemic has amplified the emotional perceptions of tourists, making their psychosocial needs more intense. Being confined to a fixed environment because of policy conditions makes tourists deviate from their habits and more or less severe in terms of psychological fluctuations. Therefore, tourists rely on technological means for psychological stress release and may prefer to experience forms of tourism with intellectual or technical attributes. Secondly, the epidemic allowed more time for researchers and tourists, guaranteeing scientific innovation in space tourism, which led to breakthroughs by major spacecraft companies during the epidemic. Finally, the willingness of people to experience space tourism is likely to grow along with the success of previous experiences. The actual Earth's environment has a series of environmental problems, virus problems, and war problems, which prompt people to desire to explore the unknown and thus realize the self-fulfilling value of life.

References

Abbas, J., Mubeen, R., Iorember, P. T., Raza, S., & Mamirkulova, G. (2021). Exploring the impact of COVID-19 on tourism: Transformational potential and implications for a sustainable recovery of the travel and leisure industry. *Current Research in Behavioral Sciences, 2*, Article 100033.

Holt, S. (2022). Virtual reality, augmented reality and mixed reality: For astronaut mental health; and space tourism, education and outreach. *Acta Astronautica.*

Huang, S., & Wang, X. (2023). COVID-19 two years on: A review of COVID-19-related empirical research in major tourism and hospitality journals. *International Journal of Contemporary Hospitality Management, 35*(2), 743–764. https://doi.org/10.1108/ijchm-03-2022-0393

Mehran, J., Olya, H., & Han, H. (2023). Psychology of space tourism marketing, technology, and sustainable development: From a literature review to an integrative framework. *Psychology & Marketing, 40*(6), 1130–1151. https://doi.org/10.1002/mar.21795

Olya, H., & Han, H. (2023). Emerging space tourism business: Uncovering customer avoidance responses and behaviours. *Journal of Vacation Marketing, 29*(3), 445–460. https://doi.org/10.1177/13567667221101408

Peng, K. L., Chan, X., & Wu, C. H. (2023). Space tourism flow generated from social media data. *Data in Brief, 48*, Article 109061.

Pickering, C., & Byrne, J. (2014). The benefits of publishing systematic quantitative literature reviews for PhD candidates and other early-career researchers. *Higher Education Research & Development, 33*(3), 534–548. https://doi.org/10.1080/07294360.2013.841651

Politis, I., Georgiadis, G., Papadopoulos, E., Fyrogenis, I., Nikolaidou, A., Kopsacheilis, A., Sdoukopoulos, A., & Verani, E. (2021). COVID-19 lockdown measures and travel behavior: The case of Thessaloniki, Greece. *Transportation Research Interdisciplinary Perspectives, 10*, 100345.

Reddy, M. V., Nica, M., & Wilkes, K. (2012). Space tourism: Research recommendations for the future of the industry and perspectives of potential participants. *Tourism Management, 33*(5), 1093–1102. https://doi.org/10.1016/j.tourman.2011.11.026

Reichel, A., Fuchs, G., & Uriely, N. (2007). Perceived risk and the non-institutionalized tourist role: The case of Israeli student ex-backpackers. *Journal of Travel Research, 46*(2), 217–226. https://doi.org/10.1177/0047287507299580

Roman, M., Kosiński, R., Bhatta, K., Niedziółka, A., & Krasnodębski, A. (2022). Virtual and space tourism as new trends in travelling at the time of the COVID-19 pandemic. *Sustainability, 14*(2), 628.

Yin, J., Bi, Y., & Ni, Y. (2022). The impact of COVID-19 on turnover intention among hotel employees: A moderated mediation model. *Journal of Hospitality and Tourism Management, 51*, 539–549.

Zaman, U., Koo, I., Abbasi, S., Raza, S. H., & Qureshi, M. G. (2022). Meet your digital twin in space? Profiling international expat's readiness for metaverse space travel, tech-savviness, COVID-19 travel anxiety, and travel fear of missing out. *Sustainability, 14*(11), 6441. https://doi.org/10.3390/su14116441

Zhang, Y., & Wang, L. (2022). Progress in space tourism studies: A systematic literature review. *Tourism Recreation Research, 47*(4), 372–383. https://doi.org/10.1080/02508281.2020.1857522

Chapter 11
Research Issues and Suggestions of Space Tourism Demand Research

1 Introduction

Space tourism began in 2001 withthe first space tourist, Dennis Tito (Crouch, 2001). The industry gained widespread recognition in 2021 due to the successful space trips of Virgin Galactic, Blue Origins, and SpaceX, respectively (Doboš, 2022; Peng et al., 2023). Even the breakthroughs in space technology, such as reusable launch systems, have reduced its high cost (Maury et al., 2020)—the prices are still extra-expensive, but there remains a high demand to book space travel (Toivonen, 2021). The profit margins gotten the equilibria of extraordinarily high pricing with low deals have made space tourism realistic and increased market appeal (Anderson Jr et al., 2023). Johnson and Martin (2016) stated that the feasibility of space travel is articulated in terms of economic and technological progress in the anticipated future. Economic conditions have allowed space tourism to enter the rapid development phase of the tourism cycle, which will attract capital and research to advance the space tourism industry (Wang et al., 2022). However, research questions have been appeared like the the economic phenomena of unexplained prices, quantities and market volatility, which could be from the foundemental differences between Earth and space tourism. We can observe that space travel suppliers are not traditional flight companies like SpaceX; space tourists were not in general that require to be wealthy with good physical and psychological health conditions. This study follows Plato's view of truth, a substantive though atypical property of statements, and supports realism about being (Hestir, 2016; Reshotko, 2022). The truth of space tourism should be told as to why some economic phenomena and behaviors are unique and still cannot be explained by existing knowledge of Earth tourism.

From an economic point of view, analysis of the uniqueness of space tourism in terms of supply, demand, and equilibrium could be a clear typology. The suppliers produce space services by innovating the production function factors, such as extending the land to the space factors. Space factors beyond the Karman line need technologies of space travel vehicles with advanced engines, new energy and fuels, as

K. Peng et al., *Space Tourism Supply Chain*, Contributions to Management Science,
https://doi.org/10.1007/978-981-95-1080-1_11

well as launch bases with concerned facilities. The supply uniqueness can be observed in the current space travel companies without existing airlines. On the consumption side, only rich and healthy people can consume space travel. The market can be at equilibrium at extremely high prices and tiny quantities with inelasticity of supply and demand curves. These differences show the research gaps between space and Earth tourism. The question is, what makes it different? One of the directions to explain it can be from Einstein's General Relativity Theory (Dolan, 2023), which advocated the spacetime construct to explain the relationship between geometry and energy–momentum tensor. Figure 1 shows the fundamental difference derived from Earth tourism traveling on zero curvature of spacetime, which we have current theories of independent space and time tourism studies. Space tourism traveling outer space beyond the Karmen relies on technologies to move along different nonzero curvature spacetime, and innovative theories can be constructed for tourism economics (Peng et al., 2023).

This study explores the different attributes of space and Earth tourism from economic perspectives and suggests research directions using three philosophical approaches. (i) The ontological approach: The Holocene to the Anthropocene (Spector & Higham, 2019) perspectives suggest that space travel helps explore the origin and civilization upgrade for human beings (Kirillova et al., 2017). Due to advanced technologies, human beings' sustainable survival alternatives have increased from current environmental conservation measures to space mobility. Space travel can be the means for the empirical study of the metaphysic philosophy of human beings; (ii) The epistemology approach: Earth and space tourism can be discussed from tourism geographies on time, space, and social physics (Hall, 2005).

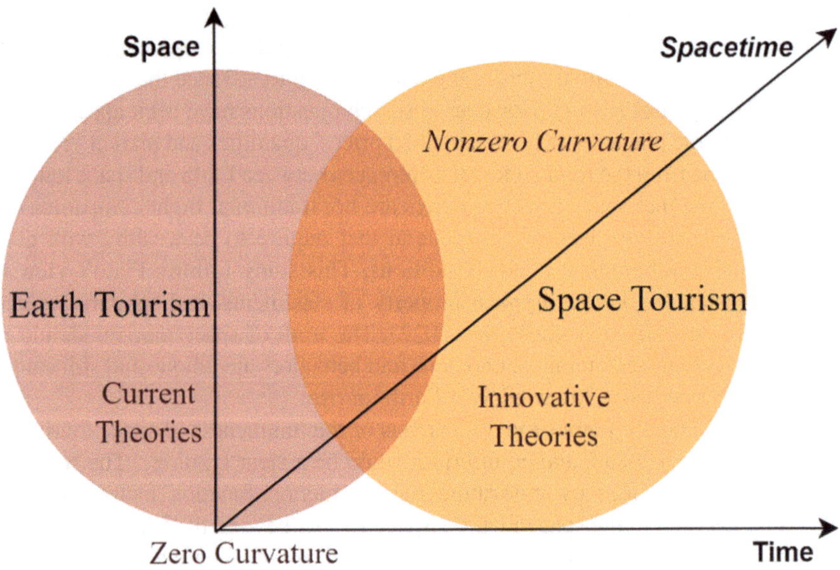

Fig. 1 Two sets of economic behaviors

Notably, space tourism is suggested to follow Einstein's general relativity theory, which advocates that spacetime scale with different curvatures to study space travel (Peng et al., 2023), (iii) The methodology approach: spacetime methods (Demkowicz et al., 2017; Miller & Haber, 2008) are awaiting for innovative research design and data analysis of space tourism because the difference between spacetime scale and independent space and time scales (An et al., 2015), not to mention the traditional Futron/Zogby survey (Ziliotto, 2010). The spacetime scale with nonzero curvature will become an innovative philosophical segmentation when we study travel behaviors in outer space, relying on space technology innovations, like SpaceX's reusable launch systems that transform the cost structure of space suppliers.

2 Space Economic Demand Behaviors

Song et al. (2023) raised nine research questions and suggested directions for the progress in tourism demand research, including demand measurement, nonlinear models, intertemporal dynamics, time series analysis, effect size, behavioral economics, micro–macro integration, and COVID-19 issues. The literature has contributed to the invariant time unit models, which can still apply to demand models of ground space tourism (Crouch, 2001), Nevertheless, space tourism demand literature has presented its uniqueness in travel experience (Chang, 2017), demography (Collins et al., 1996), demand forecasting (Crouch et al., 2009), and consumer behavior (Crouch et al., 2009; Olya & Han, 2023; Reddy et al., 2012). For example, financial and health thresholds have limited space travelers whose motivations differ from Earth tourists (Barratt, 2019; Olya & Han, 2020; Roslee, 2021). Space travelers' utilities are raised by motivations of risk-taking, challenge, curiosity, spirituality, and seeking distinction (Laing & Frost, 2019; Mehran et al., 2023; Olya & Han, 2020, 2023), especially the self-actualization based on Maslow's needs theory (Tyler, 2023). Consumer needs drive motivation for adventure is a crucial question for space tourism (Crouch, 2001). Motivation theory has innovative potential to explain why people pursue risky space travel with high vigor and persistence, like adventure tourism. Integration of motivation configuration with risk configuration was proposed to study the behavior intention of space tourism (Olya & Han, 2020). In addition, motivation models can be derived from a value-based decision-making process that determines action selection and effort expenditure on space tourism. This involves calculating, comparing, evaluating, and updating hedonic, utility, and goal values. As a result of the cost–benefit analysis, motivated individuals approach the option on sapce travel with the highest overall expected value. If existing motivation theories were to be integrated into the value-based decision-making process, then a parsimonious but highly explanatory grand theory would emerge in space tourism (Bong et al., 2023). These transformational motivation models developed from the epistemology literature could have a breakthrough to close research gaps in demand forecasting models between Earth and space tourism (Danov, 2020; Zhang & Wang, 2022).

Based on Lancaster's Characteristics Model (Wierenga, 1984), space tourism consists of unique consumer utilities—in addition to weightless experience, space scenes, knowledge enhancement, and leisure and recreation, a grand motivation of self-actualization is outstanding with the risk taker characteristics. Flying space in a Zero-G environment can never be consumed in Earth tourism choices (Crouch, 2001)—that challenge the demand attributes of traditional tourism models (Chang, 2017; Crouch et al., 2009). Space tourism's uniqueness makes the demand curve exceptionally steep and price-inelastic; consumers are price-takers with limited bargaining power over suppliers (Petri, 2021). SpaceX priced an all-civilian tour group ticket in April 2022 at approximately US$55 million. Other spacecraft also charge various expensive prices based on the orbit levels (Fig. 2) (Peng et al., 2023). Not only the high pricing taking and inelasticity attributes of space tourism demand, but tourists should also meet specific health requirements to overcome the Earth's gravity in space environments. Space tourists must overcome three to five gravity to take off and land during space travel (Braddock, 2017). The unprecedented experience of weightlessness and space landscape may also lead to psychological and cognitive dissonance (Slobodian, 2015). Hence, from a demand perspective, space tourism—especially beyond the outer space threshold (i.e., the Kármán line)—is more than an extension of Earth tourism activities. Therefore, demand forecasting and consumer behaviors measured by the curvilinear spacetime scale require cognitive adjustment to conduct demand-side research in space tourism (Peng et al., 2023).

Based on General Relativity Theory, space and Earth tourism is a status of spacetime transfer in which the curvature is changed from zero to nonzero. Earth tourism is a particular case in which the curvature of spacetime is zero. The research gap is where innovative theories are applicable to study consumer behaviors. Crouch (2001) pointed out two research gaps in space tourism demand studies: (1) The research relies on what surveyed respondents say or think they would do if presented

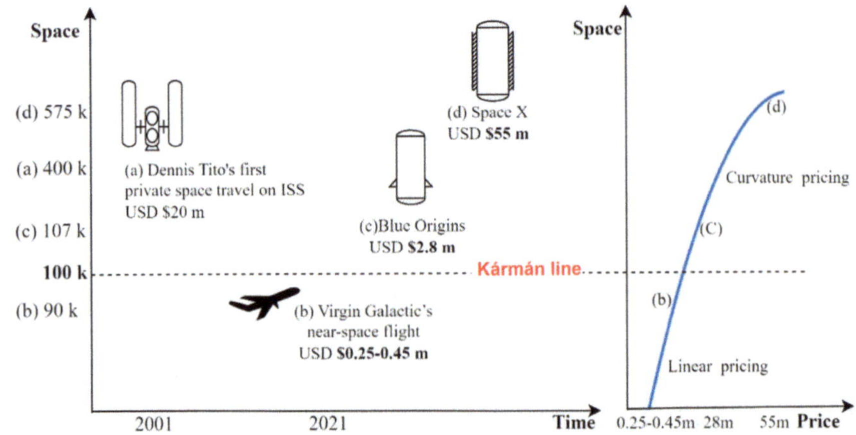

Fig. 2 Spacetime pricing at various curvatures (*Note* ISS: international space station, m: million, k: kilometer)

with the opportunity to travel into space. (2) On the issue of preparedness to pay, the survey questions leave aside the issue of the necessity for most aspiring space tourists to accumulate wealth before being able to travel into space. One of the solutions relies on motivation as a response to the perception of increased demands (Bledow, 2013), which causes a rigid demand with a greater slope of the demand curve: price inelasticity.

3 Propositions for Space and Earth Tourism

The space tourism industry is nascent (Crouch et al., 2009; Peng et al., 2023), and the very limited number of tourists has been a limit to data collection for model testing. Most space tourism studies potential customers about expectations, motivation, and behavioral intentions without real travel experience (Crouch, 2001; Zhang & Wang, 2022). Thus, it awaits more time for the space market to generate enough empirical data to test hypotheses and models of space tourism (Crouch, 2001). Even so, abnormal space tourism economic behaviors would remain unexplained without essential space attributes. Based on the literature and General Relativity Theory, we could develop the propositions refer to space attributes and observable phenomena of space economic behaviors as follows. It is also indicated in Fig. 1.

Proposition 1: Space tourism economic behaviors are determined by the nonzero curvature of spacetime.

Proposition 2: Earth tourism economic behaviors are determined by zero curvature of spacetime.

The two propositions argue that the difference between Space and Earth tourism economic behaviors are determined by the nonzero and zero curvatures of spacetime. However, Earth tourism is a special condition of zero curvature of spacetime, called flat space (Salman et al., 2022), that we have constructed current theories. The uniqueness of space tourism economic behaviors exists due to the sufficient condition of nonzero spacetime curvature. The implications are that a breakthrough cognition is needed to know and apply a spacetime scale in space tourism studies. From the supply perspective, Earth tourism uses land resources as destinations in the production function to serve and predict inbound and outbound or rural and urban tourists (Kytzia et al., 2011). However, according to the outer space law, space does not belong to any territories/lands (Peterson, 1997), which means the production factors will be redefined in space economics as described in Schumpeter's innovation theory (Williams et al., 2021). From the demand perspective, Crouch (2001) indicated that motivation theory is probably one of the crucial questions in space tourism studies. The motivation for adventure tourism evident in people's demand depends on the prices offered and product attributes. The market equilibrium depends on supply and demand, which reshapes the market structure in terms of its dynamics and competition in space tourism (Crouch et al., 2009). The invariant of economic science is still the equilibrium of supply and demand, which can be extended from zero to nonzero curvature of spacetime. Measures the utility or trade-off between various product attributes, such as trip duration, comfort, activities, safety, and other features, is still the core of demand (Crouch, 2001; Crouch et al., 2009). The production function of space attributes would still follow the cost–benefit principles. Both together attain the price equilibrium of space travel. It is expected that there will be innovative theories for space tourism beyond the Karman line. Earth tourism within the Karman line continues the incremental theories. Various studies examine the market from alternative directions and methodologies so that the weight of evidence arising from these studies is consistent, mutually supportive, and therefore sufficiently compelling (Crouch, 2001).

4 Discussions

A spacetime scale has been proposed for space tourism because it helps explain the unique behaviors distinguished from Earth tourism (Peng et al., 2023). Literature has shown a few research suggestions, but philosophical approaches are needed to construct a holistic system to study space tourism. Figure 3 shows a matrix of research gaps in the three philosophical approaches between Earth and space tourism. The areas of (1), (4), (7) are predicted to develop innovative theories of space tourism because of spacetime scale with nonzero curvature attributes that haven't been studied in current theories, areas of (3), (6), and (9) of Earth tourism. Innovative theories will be constructed when researchers adopt spacetime scale, and we could explain space tourism economic behaviors from the onotology, epistemology, and methodology approaches. Before the well constructing innovative theories, incremental theories, as indicated in the areas of (2), (5), and (8), are the bridging knowledge to guide us knowing space tourism from Earth tourism. Examples are discussed in the following sections.

Figure 3 shows that the ontological approach to metaphysics can discuss space tourism extended the philosophy of Earth tourism. Pernecky (2010) defined tourism as the mobility of people surviving and developing. The perspective can be extended from Earth tourism in three-dimensional space and independent time, as indicated

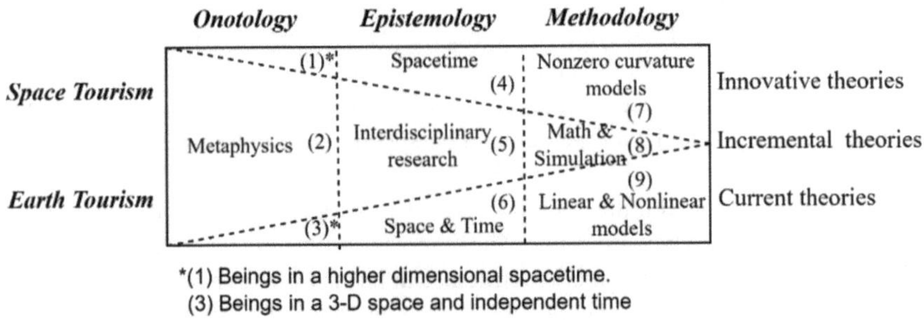

Fig. 3 The philosophical approaches in analyzing space and Earth tourism

in zone (3), to space areas of four-dimensional spacetime, shown in zone (1). Therefore, there are two directions to conduct ontological space tourism research. Firstly, we could study the scientific and societal benefits of space travel and interstellar exploration motivations for human beings. Secondly, we could apply the Kardashev Scale to discuss human civilization upscaled by space travel areas through supplier's technological advances (Namboodiripad & Nimal, 2021). Explorations of beings in a higher dimensional spactime could be extended to provide a geometrical unification of all the fundamental interactions between supply and demand based on Kaluza–Klein unified field theory and General Relativity Theory (Chyba, 1985).

From an epistemological perspective, there are differences between spacetime and independent space and time that have led to a series of research gaps in space tourism (Frost & Frost, 2022; Peng et al., 2023), which are unexplained consumer behaviors, untraditional suppliers, abnormal high pricing of market equilibrium, etc. It needs scientific cognition adjustments to construct space tourism knowledge systems from interdisciplinary fields like cosmology, physics, and math to develop innovative theories in the zone (4) and incremental theories in the zone (5), which are difficult to extend from current theories (Rahmafitria et al., 2020) of Earth tourism in the zone (6). The epistemology of spacetime has been well discussed in philosophy, physics, and math that can be applied to space economics (Dewar et al., 2022). Such a reconstruction of spacetime knowledge can be supported by empirical access in the space tourism industry because it has happened in this space era. We could observe the space industry behaviors and collect data for testing the epistemological propositions and hypotheses to construct the space economic knowledge system through interdisciplinary collaborations. For example, we can test the hypotheses from propositions 1 and 2 to tell the difference between space and Earth tourism by knowledge of Economic models and AI simulations.

Significant challenges are the methodology development of space tourism because of the nonzero curvature of spacetime when we travel into outer space, as indicated in zone (7). Many literature on space vehicle research and artificial intelligence could be approaches to studying space tourism (Figueroa, 1995). For example, Zhang et al. (2023) applied a machine-learning approach to forecasting the progression of human civilization on the Kardashev Scale, which could be used to study space tourism.

This study provides a systematic research scheme on space tourism from philosophical approaches that we can make progress in constructing a theoretical foundation. From an economic perspective, identifying the differences between space and Earth tourism can stay focused on supply, demand, and market equilibrium. Validation of the real market for public space travel and tourism will be an essential step (Crouch, 2001). However, we should have a proper research methodology to echo the spacetime reality. This study reveals the economic differences between space and Earth tourism based on the spacetime scale of GRT. The interdisciplinary analogy of physics and economics is one of the scientific approaches to explain the economic behaviors on space tourism. GRT provides strong philosophical foundations to realize that the spacetime scale makes Space Tourism.

different from Earth tourism in terms of business operations, consumer behaviors, and equilibrium adjustments. However, the spacetime concept is not intended to

Table 1 Agenda of space tourism research directions

Philosophical approoaches	Research gap	Research directions
Ontology	• Behaviors beyond the Karmen line • Beings in traditional 3-D space and independent time or higher dimensional spacetime	• Scientific and societal benefits of space travel and interstellar exploration motivations for human beings • The Kardashev Scale application of civilization upscaled by space travel areas through supplier's technological advances • Extention the Kaluza–Klein unified field theory and General Relativity Theory to beings travel in the spacetime
Epistemology	• Unexplained economic behaviors observed in space tourism • Knowledge system on space economics • Interdisciplinary theories connections and applications in space tourism	• Empirical studies on motivation theory and firm behaviors in space • Interdisciplinary analogy of spacetime scale between space tourism and Earth tourism • Applications of interdisciplinary theories from philosophy, physics, math, and cosmology
Methodology	• Linear, non-linear to curvature models	• Spacetime scale measurement and application

overthrow Earth tourism theories; instead, it aims to supplement the space tourism phenomena from a transformative economic perspective. This study aims to stimulate researchers' interest in developing a new understanding and methodology in studying space tourism with its unique attributes. Table 1 shows the summary of research directions for space tourism.

5 Conclusions

Figure 3 shows the philosophical approaches to studying space tourism from its uniqueness and the differences from Earth tourism. The ontology of being of human beings is still the central philosophy of Earth and space tourism. The epistemology differences can be told from the spacetime scale of space tourism, where the Earth tourism knowledge is based on independent space and time. Thus, the innovative theories of space tourism will be derived from the methodology that relies on the zero and nonzero spacetime curvature.

The present study addresses space tourism as a form of niche tourism; however, its uniqueness cannot be incorporated into the research stream of Earth tourism

(Abitzsch, 1996). The fundamental differences between space and Earth tourism are the curved spacetime in the universe and Earth's flat space and time, which cause different measurement scales. Albert Einstein's general relativity theory (GRT) explains that gravity is the curvature of spacetime (Angermann et al., 2022; Renn & Stachel, 2007), which tells orbital objects how to move and makes suppliers spend the most resources on space travel (Rementeria, 2022). Hawking and Ellis (1973) proposed that the spacetime scale should be adopted in space exploration fields; thus, we shall extend the spacetime scale in modeling space tourism behaviors. The anti-gravity of space travel and the need to escape the spacetime curvature (Craven, 2019) can help realize the unique behaviors of space tourism (Song et al., 2012). Recognizing gravity as the curvature of spacetime (Gomez, 2014) may close the research gaps in space tourism.

Space tourism discussed from an economic perspective has limitations of practices. Space suppliers, agencies, consumers, insurance, laws, and medicine are involved. The heavy carbon dioxide emissions from the space tourism industry toward sustainability is also a big gap to be discovered. Spacetime is a four-dimension scaler, an integrated dimension to explain economic behaviors beyond our current research cognition for space tourism. We have limited methods on spacetime methodology, as indicated in Fig. 3 zone (7). Then, the manifold mathematical approach might provide a research direction for dimensionality reduction, which can push the theoretical development from incremental innovation to innovative innovation methodologically.

References

Abitzsch, S. (1996). Prospects of space tourism. *DGLR BERICHT*, 99–112.

An, L., Tsou, M.-H., Crook, S. E., Chun, Y., Spitzberg, B., Gawron, J. M., & Gupta, D. K. (2015). Space–time analysis: Concepts, quantitative methods, and future directions. *Annals of the Association of American Geographers, 105*(5), 891–914. https://doi.org/10.1080/00045608.2015.1064510

Anderson, E. G., Jr., Parker, G. G., & Tan, B. (2023). Strategic investments for platform launch and ecosystem growth: A dynamic analysis. *Journal of Management Information Systems, 40*(3), 807–839. https://doi.org/10.1080/07421222.2023.2229125

Angermann, D., Pail, R., Seitz, F., & Hugentobler, U. (2022). The Measurement of the Earth Through Time. In *Mission Earth* (pp. 9–48). Springer. https://doi.org/10.1007/978-3-662-64106-4_2

Barratt, M. R. (2019). Physical and bioenvironmental aspects of human space flight. In *Principles of clinical medicine for space flight* (pp. 3–37). Springer. https://doi.org/10.1007/978-1-4939-9889-0_1

Bledow, R. (2013). Demand-perception and self-motivation as opponent processes: A response to Bandura and Vancouver. *Journal of Management, 39*(1), 14–26. https://doi.org/10.1177/014920631246614

Bong, M., Reeve, J., & Kim, S.-i. (2023). *Motivation science: Controversies and insights.* Oxford University Press.

Braddock, M. (2017). Ergonomic challenges for astronauts during space travel and the need for space medicine. *Journal of Ergonomics, 7*(6), 2. https://doi.org/10.4172/2165-7556.1000221

Chang, Y.-W. (2017). A preliminary examination of the relationship between consumer attitude towards space travel and the development of innovative space tourism technology. *Current Issues in Tourism, 20*(14), 1431–1453. https://doi.org/10.1080/13683500.2015.1005580

Chyba, C. F. (1985). Kaluza-Klein unified field theory and apparent four-dimensional space-time. *American Journal of Physics, 53*(9), 863–872. https://doi.org/10.1119/1.14353

Collins, P., Stockmans, R., & Maita, M. (1996). Demand for space tourism in America and Japan, and its implications for future space activities. *Advances in the Astronautical Sciences, 91*, 601–610.

Craven, M. (2019). 'Other Spaces': Constructing the Legal Architecture of a Cold War Commons and the Scientific-Technical Imaginary of Outer Space. *European Journal of International Law, 30*(2), 547–572. https://doi.org/10.1093/ejil/chz024

Crouch, G. I. (2001). The market for space tourism: Early indications. *Journal of Travel Research, 40*(2), 213–219. https://doi.org/10.1177/004728750104000212

Crouch, G. I., Devinney, T. M., Louviere, J. J., & Islam, T. (2009). Modelling consumer choice behaviour in space tourism. *Tourism Management, 30*(3), 441–454. https://doi.org/10.1016/j.tourman.2008.07.003

Danov, D. G. (2020). A review of space tourism services: supply and demand challenges. *Journal of Tourism Leisure and Hospitality, 2*(1), 29–35. https://dergipark.org.tr/en/pub/toleho/issue/53785/739977

Demkowicz, L., Gopalakrishnan, J., Nagaraj, S., & Sepulveda, P. (2017). A spacetime DPG method for the Schrodinger equation. *SIAM Journal on Numerical Analysis, 55*(4), 1740–1759. https://doi.org/10.1137/16M1099765

Dewar, N., Linnemann, N., & Read, J. (2022). The epistemology of spacetime. *Philosophy Compass, 17*(4), Article e12821. https://doi.org/10.1111/phc3.12821

Doboš, B. (2022). Tortoise the Titan: Private Entities as Geoeconomic Tools in Outer Space. *Space Policy, 60*, Article 101487. https://doi.org/10.1016/j.spacepol.2022.101487

Dolan, B. P. (2023). *Einstein's general theory of relativity: A concise introduction.* Cambridge University Press.

Figueroa, F. (1995). *Methodologies to determine forces on bones and muscles of body segments during exercise, employing compact sensors suitable for use in crowded space vehicles* NAS 1.26: 197943).

Frost, J., & Frost, W. (2022). Exploring prosocial and environmental motivations of frontier tourists: Implications for sustainable space tourism. *Journal of Sustainable Tourism, 30*(9), 2254–2270. https://doi.org/10.1080/09669582.2021.1897131

Gomez, A. L. G. (2014). Gravity is the curvature of spacetime or a quantum phenomenon. *Journal of Advances in Physics, 4*(1), 194–203.

Hall, C. M. (2005). Time, space, tourism and social physics. *Tourism Recreation Research, 30*(1), 93–98. https://doi.org/10.1080/02508281.2005.11081236

Hawking, S. W., & Ellis, G. F. R. (1973). *The large scale structure of space-time* (Vol. 1). Cambridge University Press.

Hestir, B. E. (2016). *Plato on the metaphysical foundation of meaning and truth.* Cambridge University Press.

Johnson, M. R., & Martin, D. (2016). The anticipated futures of space tourism. *Mobilities, 11*(1), 135–151. https://doi.org/10.1080/17450101.2015.1097034

Kirillova, K., Lehto, X., & Cai, L. (2017). Tourism and existential transformation: An empirical investigation. *Journal of Travel Research, 56*(5), 638–650. https://doi.org/10.1177/0047287516650277

Kytzia, S., Walz, A., & Wegmann, M. (2011). How can tourism use the land more efficiently? Model-based approach to land-use efficiency of tourist destinations. *Tourism Management, 32*(3), 629–640. https://doi.org/10.1016/j.tourman.2010.05.014

Laing, J., & Frost, W. (2019). Exploring Motivations of Potential Space Tourists. In *Space Tourism* (Vol. 25, pp. 141–161). Emerald Publishing Limited. https://doi.org/10.1108/S1571-504320190000025008

Maury, T., Loubet, P., Serrano, S. M., Gallice, A., & Sonnemann, G. (2020). Application of environmental life cycle assessment (LCA) in the space sector: State of the art. *Acta Astronautica, 170*, 122–135. https://doi.org/10.1016/j.actaastro.2020.01.035

Mehran, J., Olya, H., & Han, H. (2023). Psychology of space tourism marketing, technology, and sustainable development: From a literature review to an integrative framework. *Psychology & Marketing, 40*(6), 1130–1151. https://doi.org/10.1002/mar.21795

Miller, S. T., & Haber, R. B. (2008). A spacetime discontinuous Galerkin method for hyperbolic heat conduction. *Computer Methods in Applied Mechanics and Engineering, 198*(2), 194–209. https://doi.org/10.1016/j.cma.2008.07.016

Namboodiripad, A., & Nimal, C. (2021). Predicting the timeline for Earth-achieving Kardashev scale type 1 status. *Journal of Science and Technology, 6*, 2456–5660. https://doi.org/10.46243/jst.2021.v6.i1.pp148-152

Olya, H., & Han, H. (2023). Emerging space tourism business: Uncovering customer avoidance responses and behaviours. *Journal of Vacation Marketing, 29*(3), 445–460. https://doi.org/10.1177/13567667221101408

Olya, H. G. T., & Han, H. (2020). Antecedents of space traveller behavioral intentions. *Journal of Travel Research, 59*(3), 528–544. https://doi.org/10.1177/0047287519841714

Peng, K.-L., Hsu, C. H. C., Lin, P. M. C., & Su, M. (2023). Proposing spacetime scale for space tourism economics. *Tourism Economics, 29*(6), 1671–1678. https://doi.org/10.1177/13548166221109666

Pernecky, T. (2010). The being of tourism. *Journal of Tourism and Peace Research, 1*(1), 1–15.

Peterson, M. J. (1997). Use of analogies to develop outer space law. *International Organization, 51*(2), 245–274. https://doi.org/10.1162/002081897550357

Petri, F. (2021). Product markets: Pricing, capacity, investment, and imperfect competition. In *Microeconomics for the critical mind* (pp. 1021–1114). Berlin: Springer. https://doi.org/10.1007/978-3-030-62070-7_12

Rahmafitria, F., Pearce, P. L., Oktadiana, H., & Putro, H. P. H. (2020). Tourism planning and planning theory: Historical roots and contemporary alignment. *Tourism Management Perspectives, 35*, Article 100703. https://doi.org/10.1016/j.tmp.2020.100703

Reddy, M. V., Nica, M., & Wilkes, K. (2012). Space tourism: Research recommendations for the future of the industry and perspectives of potential participants. *Tourism Management, 33*(5), 1093–1102. https://doi.org/10.1016/j.tourman.2011.11.026

Rementeria, S. (2022). Power dynamics in the age of space commercialization. *Space Policy, 60*, Article 101472. https://doi.org/10.1016/j.spacepol.2021.101472

Renn, J., & Stachel, J. (2007). Hilbert's foundation of physics: From a theory of everything to a constituent of general relativity. In *The genesis of general relativity* (pp. 1778–1895). Springer. https://doi.org/10.1007/978-1-4020-4000-9_41

Reshotko, N. (2022). *OPINING BEAUTY ITSELF: The Ordinary Person and Plato's Forms*. State University of New York Press.

Roslee, M. F. (2021). Review of medical capability requirements for manned missions in Lunar and Martian surface base activities. *REACH, 23–24*, Article 100042. https://doi.org/10.1016/j.reach.2021.100042

Salman, M., Ali, M., & Bilal, M. (2022). Curvature inheritance symmetry in Ricci flat spacetimes. *Universe, 8*(8), 408. https://doi.org/10.3390/universe8080408

Slobodian, R. E. (2015). Selling space colonization and immortality: Psychosocial, anthropological critique of the rush to colonize Mars. *Acta Astronautica, 113*, 89–104. https://doi.org/10.1016/j.actaastro.2015.03.027

Song, H., Dwyer, L., Li, G., & Cao, Z. (2012). Tourism economics research: A review and assessment. *Annals of Tourism Research, 39*(3), 1653–1682. https://doi.org/10.1016/j.annals.2012.05.023

Song, H., Qiu, R. T. R., & Park, J. (2023). Progress in tourism demand research: Theory and empirics. *Tourism Management, 94*, Article 104655. https://doi.org/10.1016/j.tourman.2022.104655

Spector, S., & Higham, J. E. S. (2019). Space tourism in the anthropocene. *Annals of Tourism Research, 79*, Article 102772. https://doi.org/10.1016/j.annals.2019.102772
Toivonen, A. (2021). Sustainable space tourism. *Channel View Publications*. https://doi.org/10.1080/09669582.2020.1783276
Tyler, D. (2023). Self-actualization of the meaning of psychosis. *Journal of Humanistic Psychology, 63*(2), 208–214. https://doi.org/10.1177/0022167818763
Wang, L., Fu, C.-F., Wong, P. P., & Zhang, Q. (2022). The impact of tourists' perceptions of space-launch tourism: An extension of the theory of planned behavior approach. *Journal of China Tourism Research, 18*(3), 549–568. https://doi.org/10.1080/19388160.2021.1900972
Wierenga, B. (1984). Empirical test of the Lancaster characteristics model. *International Journal of Research in Marketing, 1*(4), 263–293.
Williams, A. M., Rodriguez Sanchez, I., & Škokić, V. (2021). Innovation, risk, and uncertainty: A study of tourism entrepreneurs. *Journal of Travel Research, 60*(2), 293–311. https://doi.org/10.1177/0047287519896012
Zhang, A., Yang, J., Luo, Y., & Fan, S. (2023). Forecasting the progression of human civilization on the Kardashev Scale through 2060 with a machine learning approach. *Scientific Reports, 13*(1), 11305. https://doi.org/10.1038/s41598-023-38351-y
Zhang, Y., & Wang, L. (2022). Progress in space tourism studies: A systematic literature review. *Tourism Recreation Research, 47*(4), 372–383. https://doi.org/10.1080/02508281.2020.1857522
Ziliotto, V. (2010). Relevance of the Futron/Zogby survey conclusions to the current space tourism industry. *Acta Astronautica, 66*(11–12), 1547–1552. https://doi.org/10.1016/j.actaastro.2009.08.027

MIX
Papier aus verantwortungsvollen Quellen
Paper from responsible sources
FSC® C105338

If you have any concerns about our products,
you can contact us on
ProductSafety@springernature.com

In case Publisher is established outside the EU,
the EU authorized representative is:
Springer Nature Customer Service Center GmbH
Europaplatz 3, 69115 Heidelberg, Germany

Printed by Libri Plureos GmbH
in Hamburg, Germany